Undergraduate Lecture Notes in Physics

T0254132

For further volumes:
http://www.springer.com/series/8917

Undergraduate Lecture Notes in Physics (ULNP) publishes authoritative texts covering topics throughout pure and applied physics. Each title in the series is suitable as a basis for undergraduate instruction, typically containing practice problems, worked examples, chapter summaries, and suggestions for further reading.

ULNP titles must provide at least one of the following:

- An exceptionally clear and concise treatment of a standard undergraduate subject.
- A solid undergraduate-level introduction to a graduate, advanced, or non-standard subject.
- A novel perspective or an unusual approach to teaching a subject.

ULNP especially encourages new, original, and idiosyncratic approaches to physics teaching at the undergraduate level.

The purpose of ULNP is to provide intriguing, absorbing books that will continue to be the reader's preferred reference throughout their academic career.

Ulrich Ellwanger

From the Universe to the Elementary Particles

A First Introduction to Cosmology and the Fundamental Interactions

 Springer

Dr. Ulrich Ellwanger
Department of Theoretical Physics
University of Paris-Sud
Campus d'Orsay, Bât. 210
91405 Orsay
France
e-mail: ulrich.ellwanger@th.u-psud.fr

ISSN 2192-4791
ISBN 978-3-642-44500-2
DOI 10.1007/978-3-642-24375-2
Springer Heidelberg Dordrecht London New York

e-ISSN 2192-4805
ISBN 978-3-642-24375-2 (eBook)

Cover design: eStudio Calamar, Berlin/Figueres

Printed on acid-free paper

Springer is part of Springer Science+Business Media (www.springer.com)

To my wife Gabi

Preface

It is remarkable that the fundamental laws of nature are simple. The complexity of the processes in our environment can be traced back to the fact that matter—gases, liquids, and solids—consists of an enormous number of building blocks (atoms and molecules).

Only in exceptional cases do the processes in our environment reflect the simplicity of the laws of nature. The relatively simple law of gravity allows for a description of the motion of planets or the free fall of a heavy body—but only if friction can be neglected. Even the description of the trajectory of a falling sheet of paper, where friction and other forces are important, becomes extremely complicated.

Moreover, the fundamental laws of nature seem to become progressively simpler the deeper one penetrates into the world of elementary building blocks from atoms to elementary particles. For instance, the numerous electric and magnetic phenomena can be traced back to a simple theory of electromagnetism.

However, the simplicity of such a theory reveals itself only if one employs mathematical formulations corresponding to those nature seems to use. This fact is remarkable by itself. Consequently, a certain mathematical equipment is required in order to understand the laws of nature. During recent decades this understanding has made enormous progress. We understand most of the processes in particle physics and cosmology, and manage to describe them in simple terms after making use of appropriate mathematical concepts.

The aim of this book is to present the current status of our knowledge of the laws of nature from cosmology to the elementary particles. However, we also address the numerous open questions, which often relate, interestingly enough, phenomena in cosmology to phenomena in particle physics.

The current status of our knowledge of the laws of nature encompasses four fundamental forces—gravity, electromagnetism, and the strong and the weak interactions—as well as a few elementary particles as "building blocks." Possible answers to open questions are theories that, up to now, could be neither confirmed nor disproved by experiments: amongst others, theories of the unification of three of the four fundamental forces (the exception being gravity), supersymmetry, and string theory. These theories will also be described briefly. Particle physicists and cosmologists hope to learn from experimental results within the next few years whether (or which) one of the presently speculative theories does actually describe nature.

In this text we presuppose the mathematical level of knowledge of students in natural sciences at the beginning of their studies: vector calculus, derivatives, simple differential equations and integrals. In addition, it is necessary to introduce several concepts that play an important role in cosmology and particle physics: special and general relativity, as well as classical and quantum field theory. Obviously it is impossible to describe these concepts in all detail, which would require considerably more complicated mathematical formalisms as well as a complete series of books. However, we present the essential aspects of these concepts, and many phenomena can be understood with the help of calculations that are feasible using the above mathematical equipment. In this respect, the text goes beyond the level of popular science.

The text starts with a review, beginning with the largest possible structure—the Universe—and passing by atoms and nuclei to the elementary particles, the quarks and leptons. Subsequently the corresponding concepts and physical phenomena are discussed in detail. At the end we briefly sketch the currently still speculative theories mentioned above. The text should allow a scientifically interested reader to share the fascination that accompanies penetration into the fundamental laws of nature, and possibly into a theory unifying all fundamental forces.

Finally, I would like to take this opportunity to thank several readers and Prof. D. Gromes for suggestions that helped to improve this book.

Orsay, September 2011   Ulrich Ellwanger

Contents

Chapter 1
Overview

The phenomena and the properties of the related physical objects that will be discussed in more detail in later chapters are introduced here: the Big Bang and the expansion of the Universe, and the constituents of matter. Consideration of the composition of matter leads from atoms to nuclei, protons, neutrons, and quarks. The various forms of radioactivity and the forces between elementary particles indicate the presence of new fundamental interactions, the strong and weak interactions. The presently known fundamental forces, and the elementary particles we are made of, are summarized.

1.1 The Universe

The largest possible physical object is the Universe. Since its dimensions and its dynamics far exceed our everyday experiences, and hence our imagination, it can be conceived only with the help of numbers and formulas. High powers of ten—and the ability to compute correctly with such numbers—are unavoidable in cosmology.

The visible part of the Universe consists of several hundred billion galaxies. Usually, a galaxy consists of 10^9–10^{12} stars similar to our Sun. Our Milky Way, one of the larger galaxies, contains about 3×10^{11} stars.

Galaxies are distributed approximately homogeneously throughout the Universe. However, there exist sparsely populated regions called voids, which are separated by more densely populated filaments.

Distances within and between galaxies are specified in *light years* (ly): the speed of light is given by

$$c = 299\,792\,458 \, \mathrm{m\,s^{-1}} \simeq 3 \times 10^8 \, \mathrm{m\,s^{-1}}. \tag{1.1}$$

Since a year has about 3.1536×10^7 s, a light year corresponds to about

$$1 \, \mathrm{ly} \simeq 0.9461 \times 10^{16} \, \mathrm{m} \simeq 10^{13} \, \mathrm{km}. \tag{1.2}$$

U. Ellwanger, *From the Universe to the Elementary Particles*,
Undergraduate Lecture Notes in Physics, DOI: 10.1007/978-3-642-24375-2_1,
© Springer-Verlag Berlin Heidelberg 2012

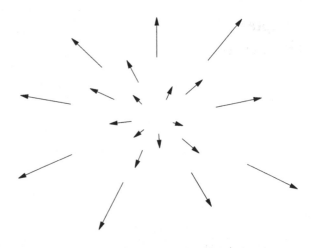

(An alternative unit is the parsec: 1 parsec = 1 pc $\simeq$ 3.262 ly, 1 kpc $\simeq$ 3.262 $\times 10^3$ ly
and 1 Mpc $\simeq$ 3.262 $\times 10^6$ ly.)

A typical size of a galaxy is 5–50 kpc or (1.5–15) $\times 10^4$ ly; the diameter of our
Milky Way is about 10^5 ly. The distances between galaxies are on the order of 10^6 ly;
the distance to the next galaxy, Andromeda, is about 2.9×10^6 ly. The largest distance
observed up to now (to a supernova, a very luminous explosion of a star) is about
10^{10} ly.

It is striking that the galaxies are veering away from us with a velocity v that
increases in proportion to their distance d:

$$v \simeq H_0 \, d, \tag{1.3}$$

where H_0 is the *Hubble constant*. This behavior is represented schematically in
Fig. 1.1, where our position is in the center, and the arrows denote the velocity
vectors of the observed galaxies.

Traditionally, velocities v of galaxies are given in kilometers per second, and
distances d—for historical reasons—in megaparsecs. For the Hubble constant one
finds

$$H_0 \simeq 70 \frac{\text{km}}{\text{s}} \times \frac{1}{\text{Mpc}}. \tag{1.4}$$

However, formula (1.3) can be applied only to sufficiently far galaxies, whose
velocities are large enough that "local" variations of up to 600 km s^{-1} can be
neglected.

In practice one often presumes the validity of the formula (1.3), and uses it to
deduce the distances of galaxies from their radial velocities with respect to the Earth.
These radial velocities are determined with the help of the Doppler effect: the wavelength of light emitted by an object moving away from us is larger ("redshifted") than
the wavelength of light emitted by an object at rest. The increase of the wavelength

is measurable, and allows the radial velocity of the object with respect to the Earth
to be determined.

Practically all other methods for the determination of distances (essentially with
the help of the luminosity, measured on Earth, of galaxies of known intrinsic bright-
ness) are in agreement with (1.3), and allow the Hubble constant to be determined.
In any case, the most important result is that the Universe is expanding.

What is the history of the Universe? Looking back in time, about 10^{10} years ago
the Universe was *compressed* and *hot*. At that time, neither the galaxies nor the
stars had formed, and the Universe was an exploding hot, dense gas. This process is
called the *Big Bang*. The stars and the galaxies formed only during the subsequent
expansion, dilution, and cooling.

The detailed study of the expansion rate of the Universe as a function of time, the
different forms of matter, the temperature, the curvature of space and space-time,
etc. is the subject of research in cosmology. The employed formalism is the theory
of general relativity, from which the gravitational forces can be derived as well (see
Chap. 3).

1.2 The Structure of Matter

Let us jump from the cosmological to the atomic scale. The orders of magnitude of
intermediate objects—which we will not discuss here—are as follows:

Systems of planets: the distance Earth–Sun is $\sim 1.5 \times 10^{11}$ m
Stars: the radius of the Sun is $\sim 7 \times 10^8$ m
Planets: the radius of the Earth is $\sim 6.4 \times 10^6$ m
Rocks, humans, ...: ~ 1 m
Grains of sand: $\sim 10^{-3}$ m
Viruses: $\sim 10^{-7}$ m
Simple molecules: $\sim 10^{-9}$ m
Atoms: $\sim 10^{-10}$ m

Observable forces in everyday life are

(a) the force of gravity,
(b) forces between bodies, the force of wind, water, combustion engines, friction
forces ....

All forces listed under (b) can be traced back to forces between atoms and molecules, and are ultimately of electric origin.

1.2.1 The Structure of Atoms

An atom consists of a cloud of *electrons*, which are negatively charged particles.
The diameter of this cloud is $\sim 10^{-10}$ m, which corresponds to the diameter of the
atom. In its center there is a positively charged *nucleus* with a diameter of some

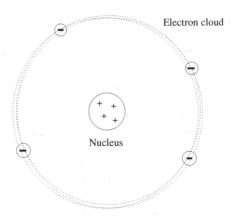

10^{-15} m (see Fig. 1.2). The angstrom, named after the Swedish physicist Anders Jonas Ångström, is defined as $1\,\text{Å} = 10^{-10}$ m (approximately the size of an atom), and the fermi (after the nuclear physicist Enrico Fermi) or femtometer as $1\,\text{fm} = 10^{-15}$ m (approximately the size of a nucleus).

The (negative) charge q_e of an electron is denoted as $q_e = -e$; the value of the elementary charge e is $e \simeq 1.60 \times 10^{-19}$ C (C stands for coulomb). The electric charge of a nucleus is always positive and a multiple of e:

$$q_{nucleus} = +Ze, \; Z = \text{integer} \tag{1.5}$$

The number of electrons of an atom is equal to Z (except for ionized atoms, where one or more electrons have been torn off). Hence, an intact atom is neutral:

$$q_{atom} = q_{nucleus} + Z q_e = Ze + Z(-e) = 0 \tag{1.6}$$

The electrons are bound to the nucleus by the electric force, since objects of opposite charge attract each other. The number of electrons—given by the charge of the nucleus—determines the chemical properties of the element. Accordingly, the charge of the nucleus defines the element:

hydrogen: $q_{nucleus} = 1e$
helium: $q_{nucleus} = 2e$
lithium: $q_{nucleus} = 3e$
. . .
uranium: $q_{nucleus} = 92e$
plutonium: $q_{nucleus} = 94e$
. . .

A detailed understanding of the structure of the electron cloud and the consequential chemical properties of atoms is possible only within the framework of quantum mechanics.

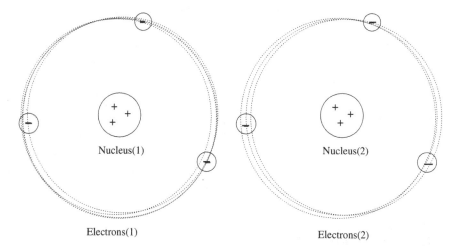

Fig. 1.3 Two atoms approaching each other

When two atoms approach each other (see Fig. 1.3), repulsive forces act between the electrons of atom 1 and the electrons of atom 2 as well as between the nucleus of atom 1 and the nucleus of atom 2, but attractive forces act between the electrons of atom 1 and the nucleus of atom 2 as well as between the electrons of atom 2 and the nucleus of atom 1.

These forces compensate each other at large distances (compared to the size of the atoms). At smaller distances of some 10^{-10} m, the balance between the forces is no longer exact, and—depending on the form of the electron clouds—the following cases are possible:

(a) The repulsion between the electrons dominates, resulting in a smallest possible distance between the atoms. For this reason one body cannot penetrate another, and our hand feels, for example, a wall.

(b) The attraction between the electrons of atom 1 and the nucleus of atom 2 and between the electrons of atom 2 and the nucleus of atom 1 dominates. In this case the atoms remain bonded, share their electron clouds, and form a molecule (see Fig. 1.4).

The forces between atoms (or molecules) are called van der Waals forces. They are complicated, but can be deduced from the (electric) forces between electrons and nuclei.

Hence, the physical phenomena in our everyday life follow from just two fundamental forces: the electric (or electromagnetic) force and gravity.

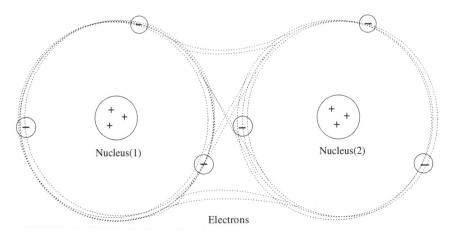

Fig. 1.4 Two atoms forming a molecule

1.3 The Structure of Nuclei

Nuclei consist of *protons* (with electric charge $q_p = +e$) and *neutrons* ($q_n = 0$). (Neutrons were discovered by J. Chadwick in 1932, for which he was awarded the Nobel prize in 1935.) Protons and neutrons are called *baryons*. The composition of a nucleus is written in the form $_Z^A X$, where X is the chemical symbol of the element, Z is the number of protons, and A the *atomic number*, which is equal to the number of baryons (the sum of protons and neutrons). For instance:

hydrogen: $_1^1 H$ (1 proton),
deuterium (chemically identical!): $_1^2 H$ (1 proton, 1 neutron),
helium: $_2^4 He$ (2 protons, 2 neutrons),
iron: $_{26}^{56} Fe$ (26 protons, 30 neutrons),
uranium: $_{92}^{238} U$ (92 protons, 146 neutrons).

Nuclei that differ only in the number of neutrons are called *isotopes* of an element.

We know that positively charged protons repel each other under the action of the electric force. Therefore, what keeps the protons (and neutrons) inside a nucleus together? Here a new force (or "interaction") enters the game, called the *strong interaction*. The strong interaction between baryons is always attractive. At small distances of a few fermi it is stronger than the electric repulsion, but it decreases rapidly at larger distances. The strong interaction is approximately independent of the nature of the baryons, that is, it is approximately identical between two protons, a proton and a neutron, and two neutrons.

The mass of a proton is $m_p \sim 1.67 \times 10^{-24}$ g. The mass of a neutron m_n is nearly the same; a neutron is just about 0.17% heavier. An electron is about 2000 times lighter: $m_e \sim 9.1 \times 10^{-28}$ g. Hence, the mass of an atom is essentially given by the mass of its nucleus.

The mass of a nucleus differs from the sum of the masses of the protons and neutrons by the *binding energy*, which contributes to the total mass, according to Einstein ($E = mc^2$),

$$m_{\text{nucleus}} = Zm_p + (A - Z)m_n - \frac{1}{c^2}E_{\text{binding}}. \tag{1.7}$$

The contribution of the binding energy is always negative, and of the order of some $10^{-2}m_p c^2$. The calculation of binding energies of various nuclei, which explain the masses of the nuclei, is one of the tasks of nuclear physics (see the exercise at the end of the chapter).

1.3.1 Radioactivity

Depending on its decomposition, a nucleus is more or less stable. An unstable nucleus can emit particles and mutate into another nucleus, which is always lighter than the original nucleus. This follows from the law of conservation of energy, which here has to be applied including the contributions of the masses to the total energy. The decay of a nucleus and the corresponding emission of particles is denoted as *radioactivity*, for the discovery and the study of which A.H. Becquerel and Marie and Pierre Curie were awarded the Nobel prize in 1903.

Let us consider the following situation: a nucleus with mass M, originally at rest, decays into n decay products with masses M_i, $i = 1 \ldots n$. After the decay, the decay products fly apart with different velocities $\vec{v}_i$, and therefore possess *kinetic energies* $E_{i\,\text{kin}} = \frac{1}{2}M_i \vec{v}_i^2$. (Here we neglect relativistic corrections, which become important only for $|\vec{v}| \sim c$, see the end of Sect. 3.1.) We denote the sum of all kinetic energies by E_{kin}. Then the law of conservation of energy reads

$$M = \sum_{i=1}^{n} M_i + \frac{1}{c^2}E_{\text{kin}}. \tag{1.8}$$

Since E_{kin} is always positive, (1.8) implies an inequality stating that the sum of the masses of the n decay products must be smaller than M. Both M and M_i have to be determined from (1.7) above, including the corresponding binding energies.

For historical reasons, the nomenclature of the radioactivities is as follows.

1.3.2 α Radiation

α denotes a helium nucleus, $\alpha = 2p2n$. The binding energy of this nucleus is particularly large, hence it is relatively light and can frequently appear as decay product. In particular heavy nuclei decay under the emission of α particles (see Fig. 1.5).

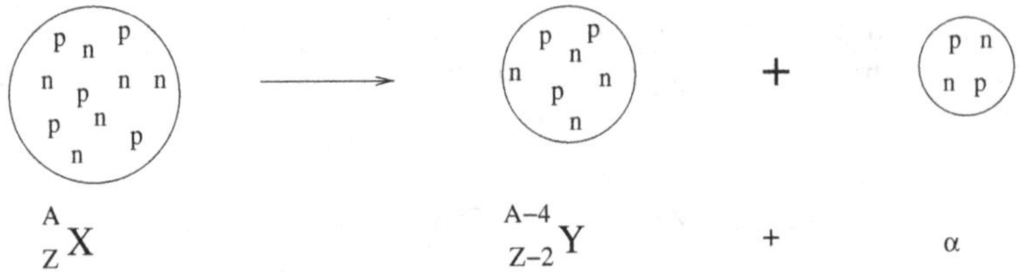

Fig. 1.5 α decay of a nucleus $^A_Z X$

Generally the α decay of a nucleus X consisting of A baryons (including Z protons) into a nucleus Y can be written as

$$^A_Z X \to ^{A-4}_{Z-2} Y + \alpha. \tag{1.9}$$

Once (1.7) is inserted for the masses of the nuclei X, Y, and α in the law of conservation of energy (1.8), one finds that the masses of the protons and neutrons cancel. From the positivity of E_{kin} one can derive a necessary condition on the binding energies of the nuclei X, Y, and α for the decay to be possible:

$$E_{binding}(X) < E_{binding}(Y) + E_{binding}(\alpha). \tag{1.10}$$

Whenever this condition is satisfied for some nucleus Y, the original nucleus X is unstable. However, due to the laws of quantum mechanics it is impossible to predict precisely the moment when the nucleus will decay; one can only measure (and attempt to compute) a half-life $\tau_{1/2}$ after which, on average, half of the nuclei have decayed.

1.3.3 β Radiation

β stands for an electron. The emission of an electron occurs mainly in the case of nuclei with more neutrons than protons (see Fig. 1.6). The emission of an electron is always accompanied by the emission of an (anti) *neutrino* $\bar{\nu}$. Neutrinos are very light neutral particles that are very difficult to detect. However, an emitted neutrino carries energy and momentum, which provides an indication that a neutrino has been emitted. W. Pauli first inferred the existence of neutrinos from the conservation of total energy and total momentum in β decays in 1930. However, their existence was proven only in 1956, for which F. Reines was awarded the Nobel prize in 1995.

After the insertion of (1.7) into (1.8), the law of conservation of energy implies

$$Zm_p + (A - Z)m_n - \frac{1}{c^2}E_{binding}(X)$$
$$= (Z + 1)m_p + (A - Z - 1)m_n$$
$$- \frac{1}{c^2}E_{binding}(Y) + m_e + \frac{1}{c^2}E_{kin}. \tag{1.11}$$

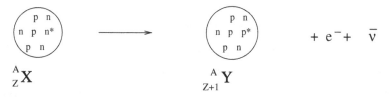

Fig. 1.6 β decay of a nucleus ^A_ZX

We see that the proton and neutron masses no longer cancel. The condition on the binding energies for the feasibility of β decay now reads

$$E_{\text{binding}}(X) - E_{\text{binding}}(Y) < (m_n - m_p - m_e) c^2. \qquad (1.12)$$

Inside a nucleus, β radiation corresponds to the conversion of a neutron into a proton, an electron, and a(n) (anti)neutrino:

$$n \rightarrow p + e^- + \bar{\nu}. \qquad (1.13)$$

In Fig. 1.6, the neutron in the original nucleus X and the proton in the newly generated nucleus Y are indicated by a star.

This process has nothing to do with the strong interaction, which—just like the electric force (or the electric interaction)—cannot change the nature of neutrons (or protons). Therefore we are dealing with a new phenomenon, the *weak interaction*. ("Weak" means "relatively rare".) A free neutron decays this way with a half-life of about 10 min (614 s). However, a neutron inside a nucleus can decay only if the binding energies of the nuclei X and Y satisfy the inequality (1.12)!

1.3.4 γ Radiation

γ stands for a *photon*. Photons are the constituents of electromagnetic radiation, i.e., X-rays, light, thermal radiation, microwaves, radio waves, etc. The equivalence of electromagnetic radiation and particles in the form of photons manifests itself in the photoelectric effect, in particular for energetic radiation such as X-rays. Amongst other things, this led to the development of quantum mechanics: according to quantum mechanics, waves of a field—such as the electromagnetic field—are equivalent to a beam of the corresponding particles, e.g. photons.

It was for the corresponding interpretation of the photoelectric effect—and not for the development of the theory of relativity—that Albert Einstein was awarded the Nobel prize in 1921.

The emission of photons occurs in the case of "excited" nuclei. Nuclei are termed excited if their binding energy is less than its maximally possible value. In that case the binding energy can increase abruptly, and the released energy is emitted in the

Fig. 1.7 Quark content of a
proton and of a neutron

Proton   Neutron

form of a photon. In this process the decomposition of the nucleus is left unchanged
in terms of protons and neutrons.

There exist other types of radioactivity, such as β^+ radiation, corresponding to the
emission of a positron; a positron is the antiparticle of an electron and has positive
electric charge. The existence of positrons was postulated in 1928 by P. Dirac (Nobel
prize in 1933), and proven in 1932 by C.D. Anderson (Nobel prize in 1936).

The constituents of nuclei, the protons and neutrons, are still not indivisible
"elementary particles":

1.4 The Structure of Baryons

Today we know that baryons are bound states of three *quarks*. There exist several
types of quarks, including the u quark with electric charge $q_u = +\frac{2}{3}e$, and the d
quark with charge $q_d = -\frac{1}{3}e$. A proton consists of two u quarks and one d quark,
and a neutron of two d quarks and one u quark, see Fig. 1.7.

This agrees with the corresponding electric charges:

$$q_p = 2 \times q_u + q_d = 2 \times \frac{2}{3}e - \frac{1}{3}e = e, \tag{1.14}$$

and

$$q_n = q_u + 2 \times q_d = \frac{2}{3}e + 2 \times \left(-\frac{1}{3}\right)e = 0. \tag{1.15}$$

Which force is responsible for the binding of quarks? This is—again—the *strong
interaction*. The attractive force between baryons can be derived from the "fundamental" force between quarks, similar to the van der Waals forces between atoms
and molecules, which follow from the electric force between nuclei and electrons.
Later we will see that there exist additional (unstable) quarks, as well as additional
baryons.

The inner structure of baryons makes it necessary to reconsider the nature of the
weak interaction: if we compare the constituents (quarks) of a neutron to those of a
proton, we find that the decay of a neutron is induced by the decay of a d quark into
a u quark:

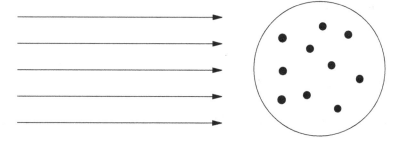

Fig. 1.8 A beam of particles hitting a cake containing hard stones

$$d \rightarrow u + e^- + \bar{\nu}. \tag{1.16}$$

This is the effect of the weak interaction at the level of quarks: it is the only interaction capable of changing the nature of quarks.

Are quarks (and the electron and the neutrino) finally elementary particles? Probably yes; until now we have not found further constituents of these particles, or determined a finite size: all we know is that they are smaller than $\sim 10^{-18}$ m. However, according to the laws of quantum mechanics, the precision Δ with which we can measure the size of an object is limited by the energy E of the particles used to bombard the object with the aim of studying its structure. At the end of Sect. 4.2 we will derive the inequality $\Delta \gtrsim \hbar c / E$ for the resolving power Δ.

A rough classical picture of this phenomenon is as follows. Let us imagine a (spherical) cake whose interior contains hard stones. In order to study the inner structure of the cake, we bombard it with a beam of particles as in Fig. 1.8.

As long as these particles carry little energy (as long as they are light and/or slow), they cannot penetrate into the cake, its interior remains concealed from us, and the cake appears to be an "elementary" object. Only if we bombard the cake with energetic particles (heavy and/or fast) do these manage to penetrate into the interior of the cake, scatter off the hard stones, and reveal the presence of the hard stones in the form of modifications of their own trajectories. (However, in the process the cake is usually destroyed.)

Similarly, the inner structure of atoms can be explored by bombarding them with, e.g., α particles (by this means, E. Rutherford discovered the atomic nucleus, for which he was awarded the Nobel prize for chemistry in 1908), and the inner structure of baryons by bombarding nuclei with electrons of higher energy such that they scatter off the constituents of the baryons, the quarks. For the detection of quarks inside baryons J.I. Friedmann, H.W. Kendall, and R.R. Taylor were awarded the Nobel prize in 1990.

However, the inner structure of objects cannot be studied with arbitrarily high precision, since only particle beams of finite energy are available. For this reason, one can determine only an upper bound on the diameter of quarks and electrons (given above), obtained with the help of particle beams of the highest currently available energy.

1.5 Preliminary Summary

Up to now we have met the following forces and particles:

Four fundamental forces (or interactions):

- the gravitational force,
- the electric (or electromagnetic) force,
- the strong interaction,
- the weak interaction.

Four elementary particles:

- the u and d quarks, which are subject to all four interactions (or feel all four forces);
- the electron, which is subject to the gravitational, the electromagnetic, and the weak (but not the strong) interactions;
- the neutrino, which feels the gravitational force (through the curvature of space-time, see Chap. 3) and the weak interaction, but neither the electric force (due to its vanishing charge) nor the strong interaction.

Elementary particles that are *not* subject to the strong interaction (such as the electron and the neutrino) are called *leptons*.

This list of elementary particles is incomplete: in addition, there exist:

(a) An antiparticle for each particle of the same mass but opposite charge. Particle–antiparticle pairs are the electron e^- and the positron e^+, and also the quarks u, d and the corresponding antiquarks $\bar{u}$, $\bar{d}$.
(b) Four additional quarks (making six quarks in all).
(c) Four additional leptons (making also six leptons in total).

However, the three elementary particles u, d, and e^- suffice to form all the different atoms that make up the matter around us.

Exercise

1.1. The dependence on A and Z of the binding energy of atomic nuclei (see (1.7)) is given to a good approximation by the Bethe–Weizsäcker formula, where the energy is expressed in units of MeV (see the appendix); $N = A - Z$ is the number of neutrons:

$$E_{\text{binding}}(A, Z) = a_v A - a_s A^{2/3} - a_c \frac{Z^2}{A^{1/3}} - a_a \frac{(N - Z)^2}{A} + \delta(N, Z). \quad (1.17)$$

The numerical values of the coefficients are $a_v = 15.56$ MeV, $a_s = 17.23$ MeV, $a_c = 0.697$ MeV and $a_a = 23.285$ MeV, and the value of $\delta(N, Z)$ is $\delta(N, Z) = 12$ MeV$/\sqrt{A}$ for N and Z even, $\delta(N, Z) = -12$ MeV$/\sqrt{A}$ for N and Z odd, $\delta(N, Z) = 0$ for $A = Z + N$ odd.

Neglect $\delta(N, Z)$, and derive a formula for $Z(A)$ that corresponds to nuclei with the largest binding energy (which are the most stable). Determine Z and find the chemical symbol for the most stable nucleus with $A = 238$.

Chapter 2
The Evolution of the Universe

The expansion of the Universe and the Big Bang are shown to be derivable from equations based on the general theory of relativity. The history of the Universe is discussed, as are various observations confirming the picture of the Big Bang, such as the abundance of light elements and the cosmological microwave background radiation. Dark matter and dark energy are introduced, and their properties and the arguments for their presence are explained. Reasons for the assumption of a so-called inflationary phase shortly before the Big Bang are given. The chapter concludes with several open questions that are the subject of today's research in cosmology.

2.1 The Expansion of the Universe in General Relativity

We have seen in the introduction that galaxies are moving away from each other, as sketched in Fig. 1.1, with a velocity v that increases in proportion to their distance d:

$$v \simeq H_0 \, d, \qquad (2.1)$$

where H_0 is the Hubble constant. Looking back in time, this implies that all matter was compressed $\sim 10^{10}$ years ago.

This phenomenon is easier to understand if we imagine a two-dimensional world instead of our three-dimensional one. A two-dimensional world corresponds to a surface, and all physical objects (and creatures) in this surface possess a width and a length, but no height. Creatures in this surface can move only inside the surface, they measure distances inside the surface and cannot even imagine a third dimension. (The mathematicians of this two-dimensional world can of course perform calculations in three-dimensional spaces; however, they have difficulties explaining to their cohabitants what this is supposed to mean.)

Let us now imagine a surface in the form of a sphere, which represents the Universe of the two-dimensional creatures. For us, this conception poses no difficulties at all;

U. Ellwanger, *From the Universe to the Elementary Particles*,
Undergraduate Lecture Notes in Physics, DOI: 10.1007/978-3-642-24375-2_2,
© Springer-Verlag Berlin Heidelberg 2012

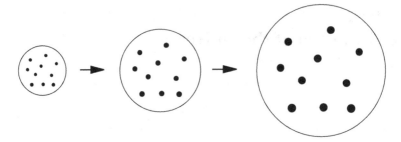

Fig. 2.1 Expanding surface of a sphere, which corresponds to a two-dimensional expanding Universe

it is, however, not conceivable for the two-dimensional creatures! In addition, we imagine that this surface is expanding, as in Fig. 2.1.

This behavior corresponds to that of our three-dimensional Universe: now, all distances between points (or galaxies) on this surface of the sphere are increasing, and the relative velocity between two points is proportional to their distance. This can be verified by a short calculation.

We introduce a dimensionless quantity $a(t)$, which is proportional to the diameter of the sphere under consideration and which increases in the course of time. $a(t)$ plays the role of a *scale factor*, i.e., all scales or lengths in the surface of the sphere are proportional to $a(t)$. We choose the convention that $a(t_0) = 1$ at the time $t = t_0$. The distance between two points measured at $t = t_0$ is denoted as Δ_0. At a later time $t > t_0$, this distance is given by

$$\Delta(t) = a(t)\Delta_0. \tag{2.2}$$

The velocity with which two points move apart from each other can be computed as follows (where $\dot{a} = da/dt$):

$$v(t) = \frac{d}{dt}\Delta(t) = \dot{a}(t)\Delta_0 = \frac{\dot{a}(t)}{a(t)}a(t)\Delta_0 = \frac{\dot{a}(t)}{a(t)}\Delta(t) = H(t)\Delta(t), \tag{2.3}$$

where

$$H(t) = \frac{\dot{a}(t)}{a(t)}. \tag{2.4}$$

Thus, $v(t)$ is indeed proportional to the distance $\Delta(t)$, but the coefficient $H(t)$ depends in general on the time t.

Here we have considered an expanding two-dimensional surface whose curvature is the same everywhere. There exist more two-dimensional surfaces with this property, which is denoted as "homogeneity": the flat plane and a surface in the form of a saddle. Equations 2.3 and 2.4 are valid in all these cases, as well as for our three-dimensional Universe: a warped three-dimensional space (or an expanding three-dimensional space) is just as unthinkable for us as a two-dimensional space with these properties for the two-dimensional creatures. Nevertheless, the calculation (2.3) above still implies a relation of the form (2.1); it suffices to replace t by $t = t_{today}$ everywhere in (2.3).

Today's experiments allow us even to measure the time dependence of $H(t)$: for very distant supernovae, the ratio of their velocity to their distance is not exactly constant, since their light was emitted very long ago and the value of $H(t)$ at that time was not exactly the same as today. Later we will discuss implications of these measurements.

We should note that an increasing scale factor $a(t)$ does *not* imply that objects in the Universe (such as stars and galaxies) are expanding: the diameter of such objects is determined by the compensation of the forces that act on their constituents (e.g. the gravitational and centrifugal forces acting on stars in galaxies). As long as these forces remain the same, the diameters of objects remain unaffected by the expansion of the Universe.

The time dependence of $a(t)$—and accordingly that of $H(t)$—can be computed in the framework of general relativity. In general relativity, space (and even space-time, see Chap. 3) is generally considered as warped (or curved). The detailed form of a warped space is determined by the distances between points everywhere in the space. The mathematical quantity describing these distances is denoted as the *metric*, which we will discuss in more detail in Chap. 3. For homogeneous spaces, the metric does not depend on the position and is completely determined by the scale factor $a(t)$ introduced above.

Einstein used the metric for the description of warped spaces in the theory of general relativity, and proposed equations that determine the metric in terms of matter (and energy) distributed in space [1].

If a homogeneous Universe is assumed, all kinds of matter (galaxies, stars, dust, atoms, elementary particles) can be considered as a homogeneous gas. In general, this gas consists of several components, but it is completely specified by its *matter density* ϱ (measured in kg/m^3) and its pressure p. For a homogeneous gas, these quantities do not depend on the position, but solely on the time t.

In general, one has to distinguish the following kinds of matter and energy:

(a) Bodies moving slowly compared to the speed of light, such as galaxies, stars, dust, and (massive and not too energetic) elementary particles. The contribution of these bodies to the density ϱ is denoted as ϱ_{nr} (where "nr" indicates non-relativistic objects with velocities $v \ll c$). The contribution of these objects to the "pressure of the Universe" is negligibly small.

(b) Massless (or light and energetic) particles that move at (or near) the speed of light provide a contribution ϱ_r to the density as well as a contribution p to the pressure, where p and ϱ_r are related by $p \sim \frac{1}{3}\varrho_r c^2$.

(c) Constant fields (see Chap. 4, "Field Theory", and Chap. 7, "The Weak Interaction") can generate a potential energy (density), which is called the *dark energy* or the *cosmological constant* Λ and measured in units of $(kg\,m^2/s^2)/m^3 = kg/(m\,s^2)$.

The Einstein equations lead to two equations for the time derivatives of $a(t)$, depending on $\varrho = \varrho_{nr} + \varrho_r$, p, and Λ. It is convenient to define a gravitational constant κ related to Newton's constant G:

$$\kappa = \frac{8\pi G}{c^2} \simeq 1.866 \times 10^{-26}\,\mathrm{m\,kg^{-1}}. \tag{2.5}$$

Using the standard definitions $\dot{a} = da/dt$ and $\ddot{a} = d^2a/dt^2$, these equations are of the following form (under the assumption that the homogeneous Universe is *not* warped, which agrees best with the observations):

$$3\frac{\dot{a}^2}{a^2} = \kappa\left(\Lambda + \varrho(t)c^2\right), \tag{2.6}$$

$$2\frac{\ddot{a}}{a} + \frac{\dot{a}^2}{a^2} = \kappa\left(\Lambda - p(t)\right). \tag{2.7}$$

These equations are also denoted as the Friedmann–Robertson–Walker equations (see, e.g., A. Friedmann in [2]).

In the present Universe the contribution of the pressure $p(t)$ to (2.7) is negligible. If one neglects Λ as well, the right-hand side of (2.7) vanishes. The left-hand side can be written in terms of the function $H(t)$ defined in (2.4), and we obtain

$$2\dot{H}(t) + 3H^2(t) = 0. \tag{2.8}$$

The general solution of this equation is given by $H(t) = 2/\big(3(t - \bar{t})\big)$, $\bar{t}$ arbitrary, and it is convenient to choose $\bar{t} = 0$ for the "origin of time". Then we get

$$H(t) = \frac{2}{3t}. \tag{2.9}$$

Now, $a(t)$ can be determined from (2.4):

$$a(t) = a_0 t^{\frac{2}{3}}, \tag{2.10}$$

where a_0 is an arbitrary constant. Consequently $a(t)$ increases with t, corresponding to an expanding Universe.

$\varrho(t)$ and $p(t)$ always satisfy a relation that follows from the conservation of energy (or from a combination of (2.6) and (2.7)):

$$\dot{\varrho}(t) = -3\frac{\dot{a}}{a}\left(\varrho(t) + p(t)/c^2\right). \tag{2.11}$$

In the case $p(t)=0$, it follows that

$$\varrho(t) = \frac{\varrho_0}{a^3}, \tag{2.12}$$

where ϱ_0 is a free constant.

Assuming $\Lambda = 0$, (2.6) and (2.10) or (2.12) now allow the (complete) matter density $\varrho(t)$ to be determined:

$$\varrho(t) = \frac{4}{3\kappa c^2 t^2} = \frac{\varrho_0}{a_0^3 t^2}. \tag{2.13}$$

Accordingly the matter density decreases, which is understandable as the volume of the Universe increases as a^3. (Eq. (2.13) can be considered as an equation for a_0 for a given constant ϱ_0: $a_0^3 = \frac{3}{4}\kappa\varrho_0 c^2$.)

2.2 The History of the Universe

As an apparent consequence of (2.13), the matter density $\varrho(t)$ was very high in the early Universe (for small t). According to the laws of thermodynamics, the temperature increases in a compressed gas. Thus, the temperature in the early Universe was very high. A high temperature of a gas implies high average velocities of its components. Collisions between these components can break them up into their subcomponents: with increasing temperature and density first molecules into atoms, then atoms into electrons and nuclei, then nuclei into baryons (protons and neutrons), and finally even baryons into quarks.

If the evolution of the Universe is described by (2.6) and (2.7), all this happened in reverse order: at the beginning, the Universe was extremely dense and hot, filled with elementary particles such as quarks and electrons. (As long as the average velocities of these particles are close to the speed of light, they contribute to the pressure $p(t) \sim \frac{1}{3}\varrho_r c^2$. Then, (2.6) and (2.7) imply—under the assumption $\Lambda \sim 0$— that $a(t) \sim a_0\sqrt{t}$ instead of (2.10) during this early stage.) This Universe sort of exploded: it expanded very rapidly, whereupon its temperature and density decreased. This process is known as the "Big Bang". In the course of time, the baryons, nuclei, atoms, molecules, and ultimately the stars and galaxies formed.

Using (2.6) and (2.7), the laws of thermodynamics (which allow the temperature to be determined as a function of the density and the pressure), and the known interactions between quarks, baryons, nuclei, and electrons, the history of the Universe can be reconstructed quite precisely, and observable consequences of this scenario can be predicted.

During the first 10^{-12} seconds the temperature was so high (above 10^{15}°C) that the processes that occurred depended on the properties of very massive— still unknown—elementary particles. (Very massive elementary particles can not yet have been produced at present accelerators, see Chap. 8.) This period is the subject of ongoing research in particle physics and cosmology. One topic of particular interest is the origin of the disequilibrium between matter and antimatter (the present Universe contains practically no antimatter); for the generation of this disequilibrium, processes that occur at such temperatures can play an important role.

After about 10^{-6} seconds (at a temperature of about 10^{12}°C), the quarks formed protons and neutrons.

After about 10 seconds (at a temperature of 10^9-10^{10}°C), the protons and neutrons formed the nuclei of light elements such as deuterium, helium, the isotope helium-3, and lithium. (Hydrogen, whose nucleus consists of just one proton, remained the most frequent element after this stage.)

After about 4×10^5 years (at a temperature of about 3000°C) the atoms were built out of nuclei and electrons.

After about 10^8 years (at a temperature of about 30 kelvin (30 K)) the stars and galaxies formed. In the interior of these stars, and during the first explosions of supernovae, nuclei of heavy elements such as iron and uranium were generated.

After about 10^{10} years (at a temperature of about 6 K) the solar system formed. It contains heavy elements, mainly in the planets, which were produced during the previous period.

Today the Universe has an age of about 1.4×10^{10} years, and has cooled to a temperature of 2.73 K.

Are there implications of this history of the Universe that are observable today?

The first of the processes described above leading to a verifiable prediction is the formation of the light elements. The relative abundance of protons to neutrons at that time (about 7:1) is calculable, and allows the determination of the relative abundance of the light elements hydrogen, helium, lithium, and their isotopes. The results of these calculations agree well with the measurements of the relative contributions of these elements ($\sim 75\%$ hydrogen, $\sim 24\%$ helium, see Exercise 2.2) to the density of gaseous clouds that originated during the primitive Universe.

Until the formation of atoms by nuclei and electrons, the constituents of the gas filling the Universe carried electric charges; subsequently the electric charges of the nuclei and electrons became neutralized in the atoms. The high temperature of the gas corresponds to chaotic motions at high velocities, and high accelerations generated by collisions. Under such circumstances, charged particles at temperatures above about 1000°C emit electromagnetic radiation corresponding to visible light. (A flame is a gas at a temperature high enough that electrons are ripped off atoms as a result of violent collisions. This gas contains ionized atoms and free electrons; it is called a *plasma*. When ionized atoms capture electrons, light is emitted.)

Hence, the Universe was full of electromagnetic radiation interacting with charged particles (i.e., being emitted, absorbed, or scattered), until the electrons and nuclei combined to neutral atoms. After the formation of (neutral) atoms the production of electromagnetic radiation ceased.

What became of the light originating from this period? A large fraction has not been absorbed up to now, and is still present in today's Universe. However, between the moment this light was produced and now, the Universe has expanded by about a factor of 1000. Simultaneously, the wavelength of the radiation in the Universe has been stretched by the same amount. Originally, this wavelength corresponded to $\lambda_{light} \sim 7 \times 10^{-7}$ m; accordingly it corresponds to microwave radiation today. It is also known as cosmic background radiation, and glares uniformly from all directions in the sky. The dependence of the intensity of the radiation on wavelength agrees with the calculations to a relative precision of 10^{-5}, and corresponds to the electromagnetic radiation of a body of a temperature of 2.73 K. For this reason we can consider this temperature as the temperature of the Universe: every object in empty space (sufficiently far from radiating stars and galaxies) will cool down to this temperature.

The cosmic background radiation following from the theory of the Big Bang was predicted, amongst others, by R. Dicke and G. Gamow, and detected by A.A. Penzias and R.W. Wilson in 1964–1965, for which they were awarded the Nobel prize in 1978.

The genesis of stars and galaxies after about 10^8 years took place under the action of gravity, which could play a role only after the motions generated by temperature had sufficiently died away. The formation of lumps of matter under the influence of gravity required, however, small density fluctuations in the gas at that time. We can deduce, from the order of magnitude of the density variations at that time, the density variations during the much earlier epoch when the atoms formed. These density variations of electrons and nuclei manifest themselves, in turn, in the form of inhomogeneities of the radiation (the light) at that time, which leads to inhomogeneities of the presently observed cosmic background radiation. This implies that the intensity of the cosmic background radiation observed today should depend weakly on the direction in the sky; the predicted relative intensity variations $\Delta I / I$ of the order of 10^{-5} were first detected in 1992 by instruments placed on the COBE (Cosmic Background Explorer) satellite, for which the Nobel prize was awarded to J.C. Mather and G.F. Smoot in 2006.

Hence, the theory of the Big Bang—at least as of 10^{-6} seconds after the origin of the Universe—has been confirmed by several observations and measurements that are based on very different physical phenomena.

2.3 Dark Matter and Dark Energy

Let us return to the Friedmann–Robertson–Walker equations (2.6) and (2.7), from which we will draw additional conclusions. The solutions (2.9) for $H(t)$, (2.10) for $a(t)$, and (2.13) for $\varrho(t)$ have been derived under the assumption that the contributions from the pressure $p(t)$ and the cosmological constant Λ can be neglected in (2.6) and (2.7). Even though the pressure played a role in the early Universe, the solution (2.9) allows quite a precise estimate of the age of today's Universe.

The age of the Universe t_{today} can be determined from the present value $H_0 \simeq$ 70 km/s $\times$ 1/Mpc of the Hubble constant. After converting megaparsecs into kilometers we obtain from (2.9)

$$t_{\text{today}} \equiv t_0 \sim 1.4 \times 10^{10} \text{ years}, \tag{2.14}$$

which also corresponds approximately to the age of the oldest stars and galaxies.

Fig. 2.2 Radius r and
rotational velocity $v(r)$ of a
star rotating around the
center of a galaxy

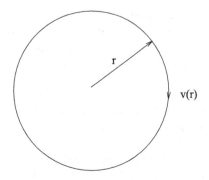

Then we obtain for the matter density $\varrho(t_0)$ from (2.13)

$$\varrho(t_0) \sim 2 \times 10^{-27} \, \text{kg m}^{-3}. \tag{2.15}$$

This value can be compared to the density of galaxies and intergalactic dust. The corresponding density of known matter ϱ_{known} is smaller than the value (2.15):

$$\varrho_{\text{known}} \sim \frac{\varrho(t_0)}{6}. \tag{2.16}$$

This means that, besides the known matter, there should exist an unknown form of "dark matter" ("dark" since, evidently, it does not emit light). The contribution of dark matter to the total matter density seems to be about five times the contribution of known matter.

At this point we should discuss a phenomenon related to the dynamics of stars inside galaxies: the nearly circular motion of stars around the center of galaxies is caused by the gravitational attraction between the stars. From the known form of the gravitational force we can compute the rotational velocity $v(r)$ of a star (sketched in Fig. 2.2), which depends on its distance r to the center of the galaxy and the mass $M(r)$ inside a fictitious sphere with radius r (G is Newton's gravitational constant):

$$v^2(r) = \frac{GM(r)}{r}. \tag{2.17}$$

In practice we can measure the rotational velocities $v(r)$ of stars at different distances r to the galactic center for a large number of galaxies and estimate $M(r)$. Surprisingly, the observations do not agree with (2.17): either the measured values of $v(r)$ are systematically too large, or the estimates of $M(r)$ are systematically too small! (Notably for large r, where the density of stars decreases and where $M(r)$ should hardly increase with r, $v(r)$ does *not* decrease as $1/\sqrt{r}$, but remains approximately constant.) This discrepancy suggested, already before cosmology, the existence of additional dark (invisible) matter, which contributes to $M(r)$ and thus to the attractive gravitational force of galaxies. Hence, there exist two independent reasons for the hypothesis of dark matter.

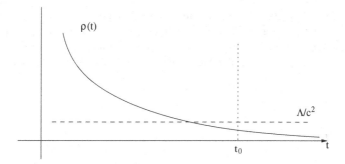

Fig. 2.3 Schematic time dependences of $\varrho(t)$ ($\sim 1/t^2$) and Λ (constant)

In recent years, very distant supernova explosions whose light was emitted a very long time ago have been successfully observed [5–8]. Through measurements of their radial velocities, with the help of the Doppler effect, and their distances, with the help of the known luminosity of such supernova explosions, the time dependence of $H(t)$ was determined for the first time, allowing a comparison to solutions of the Friedmann–Robertson–Walker equations. It appeared that $\dot{H}(t)$ is somewhat larger than expected from the above solution, which was derived under the assumption $\Lambda = 0$. The measured value of $\dot{H}(t)$ is compatible only with a positive value of Λ (a "dark energy") in (2.6) and (2.7):

$$\Lambda \sim 4 \times 10^{-10}\,\mathrm{kg\,s^{-2}\,m^{-1}}. \tag{2.18}$$

The Nobel prize 2011 was awarded to Saul Perlmutter, Adam Riess and Brian Schmidt for this observation.

First of all we have to verify whether this value invalidates the results obtained above under the assumption $\Lambda = 0$. Fortunately this is not the case. Comparing the two terms on the right-hand side of (2.6) one finds that, on the one hand, both are of the same order today:

$$\Lambda \sim 2 \times \varrho(t_0)c^2. \tag{2.19}$$

However, the time dependences of the two terms are very different: $\varrho(t)$ behaves as $1/t^2$, but Λ can be considered as constant (see Fig. 2.3).

Correspondingly, at earlier times, i.e., for $t \ll t_0$, $\varrho(t)$ was much larger than Λ/c^2, and Λ was numerically negligible. (In fact, this is also true for (2.7), as we can verify explicitly for $p(t)=0$ and inserting the above solution for $a(t)$.) For this reason, Λ has had an impact on the evolution of the Universe only in recent times; corresponding small corrections have already been taken into account in the value (2.14) for the age of the Universe.

Because of the different time dependences of $\varrho(t)$ and Λ, it appears a remarkable coincidence that—as noted in (2.19)—$\varrho(t_0)c^2$ and Λ are of the same order today. Accordingly, we live in a kind of transition period: in the (still very far) future the evolution of the Universe will be determined nearly exclusively by the Λ terms in (2.6) and (2.7), whereupon $a(t)$ will increase exponentially with t, in contrast to (2.10) (see the next section). Then, the Universe becomes infinitely large, empty and cold. However, before that (in about five billion years) our Sun will balloon to a red giant star.

2.4 Inflation

In fact, the practically homogeneous distribution of galaxies and the cosmic background radiation within the presently observable part of the Universe poses a puzzle. The homogeneous distribution of galaxies and the cosmic background radiation implies that the hot and compressed gas of elementary particles that the Universe consisted of at its beginning was also distributed very uniformly.

However, a gas can spread uniformly only if its constituents can flow back and forth. Independently of their precise nature, the flow velocity of these constituents is always limited by the speed of light. Therefore these constituents can move at most a distance $\Delta d = c\Delta t$ within a given time interval Δt.

At the beginning of the Universe, during the period of the Big Bang, this distance was not large enough in order to encompass all of the Universe observable today. (Even light needs billions of years to cross the present Universe.) Since the gas at that time could not spread uniformly within the presently observable region of the Universe, the presently almost homogeneous distribution of galaxies and the cosmic background radiation is a paradox at first sight.

In order to resolve this paradox, so-called *inflation* was invented [9, 10], which corresponds to the following behavior of the early Universe. At first we content ourselves with the fact that the original gas can spread uniformly, within the time interval Δt, only within distances $\Delta d = c\Delta t$, where Δd is very much smaller than the present Universe. Subsequently we can make use of the behavior of the solutions of the Friedmann–Robertson–Walker equations (2.6) and (2.7) in the case where the parameter Λ is much larger than $\varrho(t)$ and $p(t)$. Then, the time dependence of the scale factor $a(t)$ is no longer given by (2.10) but—as can be easily checked—by

$$a(t) = a_0 e^{\sqrt{\kappa \Lambda/3}\, t}. \qquad (2.20)$$

This implies an extremely fast—exponentially growing—expansion of the Universe, much faster than described by (2.10) before. (Such a Universe is known as a *de Sitter Universe*.)

Thereby, the distance Δd inflates to $e^{\sqrt{\kappa \Lambda/3}\, t}$ times its original value as well! This process is called *inflation*. If the period of inflation continued for a time interval Δt with $\sqrt{\kappa \Lambda/3}\,\Delta t \gtrsim 60$, the initial distance Δd within which the original gas was uniformly distributed expanded sufficiently in order to encompass the presently visible part of the Universe.

On the one hand, this process would explain the present homogeneous distributions of galaxies and the cosmic background radiation. However, we know that the Universe has no longer been expanding exponentially for about 1.4×10^{10} years, otherwise all previous results would no longer be valid. Hence we have to assume that the inflationary phase ceased after Δd had sufficiently inflated. This implies that the parameter Λ had to shrink from a relatively large value to the relatively small value given by (2.18).

Thus we have to understand how the parameter Λ can change with time. This is comprehensible in the context of field theory as used in particle physics: in field theory one obtains contributions to the potential energy that, in turn, depend on the presence of a constant field. The minimization of such a potential energy as a function of the so-called Higgs field will play an important role in Sect. 7.3 on the weak interaction. In cosmology, this potential energy acts precisely as a parameter Λ in the Friedmann–Robertson–Walker equations (2.6) and (2.7). Once a field varies in time (since it always tries to minimize its potential energy), the potential energy can decrease from a large value to a small value. (We will come back to this behavior at the end of Sect. 7.3.) Such a mechanism explains the end of an inflationary period, and now all results obtained previously are to be interpreted in the era "after the end of inflation".

Actually, such an end of an inflationary epoch through the variation of a field has additional consequences: before a field settles down to a new value (which minimizes the potential energy), it wiggles a little bit and radiates energy in the form of particles, which generates—albeit tiny—density fluctuations. This fits well with the considerations at the end of Sect. 2.2, in which small density fluctuations of the original matter are a necessary condition for the possibility that matter lumps together to form stars and galaxies under the action of gravity.

This also leads to relative intensity variations $\Delta I/I$—depending on the direction in the sky—of the cosmic background radiation observed today: if we measure the intensities of the cosmic background radiation in different directions in the sky, separated by an angle θ, they differ by about 0.001%. In addition we can now compute how this difference depends, on average, on the angle θ. This θ dependence of the intensity variations has been measured by instruments on the WMAP satellite (see the internet address given in the appendix), and it agrees well with the inflationary model.

2.5 Summary and Open Questions

The standard model of cosmology including the Big Bang has led to various predictions that agree very well with measured observables: the temperature and the (tiny, but measurable) variations of the cosmic background radiation as well as the relative abundance of light elements; the most relevant observation is, of course, that the radial velocity of galaxies increases with their distance. However, several questions still remain open.

(a) What does dark matter consist of? Practically all forms of known matter (e.g., cold, invisible stars, dust, or gas) are excluded, since they would absorb too much light if their abundance or density should explain all of dark matter. One possibility would be a new species of elementary particles (so-called *WIMPs*, weakly interacting massive particles), which should be: (1) neutral, in order not to absorb too much light; (2) stable, in order not to have decayed yet; (3) relatively heavy such that their average velocity is much smaller than the speed of light—otherwise they would contribute to the pressure term $p(t)$ in (2.7) (which is not observed), and $M(r)$ in (2.17) could not depend on r in the observed way. None of the known elementary particles satisfies all these conditions! We believe for this reason, amongst others, that there exist new elementary particles still to discover, which are the constituents of dark matter (see also Sect. 12.2 on supersymmetry).

(b) What is the origin of the dark energy (or the cosmological constant)? As we already mentioned above, its present numerical value—of the same order as the matter density $\varrho(t_0)$—is a coincidence that is difficult to explain. A real problem appears in the context of field theory mentioned above: in this theory we obtain contributions to the potential energy (or "vacuum energy") that correspond to the cosmological constant but which exceed its value given in (2.18) by many orders of magnitude (by a factor 10^{54} in the framework of weak interaction; see the end of Sect. 7.3). The fact that a large value of Λ was actually desirable during an inflationary epoch does not facilitate an explanation of its relatively small value today. Either we have not yet understood an essential aspect of the relevant theory, or there are many different contributions to Λ that cancel nearly exactly after the end of the inflationary epoch. However, at present nobody is aware of a mechanism that would lead to such a compensation of different contributions; this problem is called the "problem of the cosmological constant".

(c) Normally we should assume that, after the Big Bang, the Universe contains as many particles as antiparticles. However, the observable part of the Universe contains practically no antimatter, just "ordinary" matter. That is, evidently processes occurred that break the matter–antimatter symmetry. Indeed, we have already observed a violation of this symmetry in decays of certain particles (see the so-called CP violation in Sect. 7.4). However, at present it is not clear whether this symmetry violation suffices to explain the present disequilibrium between matter and antimatter; to this end we need a better understanding of processes that took place at a time before 10^{-12} s (at a temperature above $10^{15}\,^\circ$C).

(d) Did the Universe really undergo an inflationary epoch? If yes, what precisely did it look like? (See also the end of Sect. 7.3.) Which field, or which potential energy, was responsible for it? Is it really true that an oscillating field at the end of an inflationary epoch is responsible for the density fluctuations at the origin of the formation of stars and galaxies? In order to learn more about this inflationary epoch, a better knowledge of the angular dependence of the intensity variations of the cosmic background radiation would be very helpful. We hope to gain such information with the help of instruments placed on the Planck satellite, which was launched in 2009.

(e) What was the origin of the Big Bang? What happened at times before 10^{-12} s, or even before $t = 0$? Equations (2.6) and (2.7) can no longer be valid in the limit $t \to 0$, and the answers to these questions depend on how these equations are modified. Different theories beyond the Einstein equations lead to different modifications, but nobody knows at present whether—or which of—such theories (amongst others theories in which space-time is higher dimensional, see Sect. 12.3) are realistic.

Exercises

2.1. Solve both Friedmann–Robertson–Walker equations (2.6) and (2.7) for $\Lambda = 0$, $p(t) = w\varrho(t)c^2$ for arbitrary constants w. ($w = 0$ corresponds to a Universe dominated by massive particles and $w = 1/3$ to a Universe dominated by massless particles. Show that $w = -1$ is equivalent to $p(t) = \varrho(t) = 0$, $\Lambda \neq 0$.)

2.2. Assume that, before the formation of light nuclei, the Universe contains free protons and neutrons with a ratio 7:1. Assume, in addition, that only the particularly stable helium nuclei ${}_2^4\mathrm{He}$ form, but free protons ${}_1^1\mathrm{H}$ (hydrogen nuclei) remain as well. Derive the ratio of densities $\varrho_\mathrm{H} : \varrho_\mathrm{He}$ after the formation of light nuclei.

Chapter 3
Elements of the Theory of Relativity

This chapter introduces the concepts relevant to the special and general theories of relativity. In the theory of special relativity, measured time intervals are, in general, distinct for different observers. In this framework, space and time can be described in the form of a particular four-dimensional space–time. The new expressions for the kinetic energy and the momentum imply that no object can move faster than the speed of light. In the general theory of relativity, the four-dimensional space–time can be curved, and this curvature in the proximity of stars or planets generates the gravitational force. It is shown that, under certain conditions, the very same curvature turns compact stars into black holes.

3.1 The Special Theory of Relativity

In everyday life it is considered self-evident that time is "absolute", the same quantity for everyone independent of their position and velocity. This conception corresponds to Fig. 3.1, where vertical dotted lines denote an absolute time, the horizontal line our time-independent position, and the inclined line the motion of an astronaut at constant velocity.

The points A and B in Fig. 3.1 denote two events that take place at a given time at a certain position chosen inside the spaceship of the astronaut.

For us "here" the two events take place at different positions; the spatial distance $\Delta x_{\text{us}}^{\text{AB}}$ between the two events can easily be read off Fig. 3.1 along the vertical x-axis. For the astronaut the spatial distance $\Delta x_{\text{astr}}^{\text{AB}}$ between the events vanishes; his spatial coordinate system moves relative to us at constant velocity v_x, and in this coordinate system the two events take place at the same position. We consider it as evident, however, that the chronological interval Δt^{AB} between the events is the same for us and for the astronaut.

Written in formulas, the relations between the spatial and chronological intervals measured by us and the astronaut read

$$\Delta x_{\text{astr}}^{\text{AB}} = \Delta x_{\text{us}}^{\text{AB}} - v_x \Delta t_{\text{us}}^{\text{AB}}, \quad \Delta t_{\text{astr}}^{\text{AB}} = \Delta t_{\text{us}}^{\text{AB}}; \tag{3.1}$$

U. Ellwanger, *From the Universe to the Elementary Particles*,
Undergraduate Lecture Notes in Physics, DOI: 10.1007/978-3-642-24375-2_3,
© Springer-Verlag Berlin Heidelberg 2012

Fig. 3.1 x–t–diagram of an
astronaut moving at constant
velocity, and a person at rest
"Here"

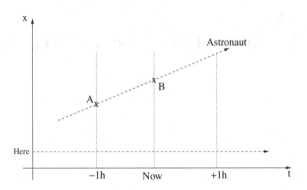

these are the transformation laws according to Newtonian mechanics between two
coordinate systems moving at constant relative velocity v_x. If one substitutes for v_x
the velocity of the space ship inside which the events A and B took place,

$$v_x = \frac{\Delta x_{\text{us}}^{\text{AB}}}{\Delta t_{\text{us}}^{\text{AB}}}, \tag{3.2}$$

one finds indeed $\Delta x_{\text{astr}}^{\text{AB}} = 0$.

In special relativity the transformation laws (3.1) are no longer exactly the same,
but valid only approximately as long as the velocity v_x is small compared to the
speed of light c. In particular the chronological intervals Δt^{AB} measured by us and
by the astronaut between the *same* events A and B are not identical!

In fact there exists no fundamental principle which requires the existence of an
absolute time (and hence $\Delta t_{\text{astr}}^{\text{AB}} = \Delta t_{\text{us}}^{\text{AB}}$), apart from our force of habit and our
restricted imagination based on experiences at velocities much smaller than c.

In order to grasp the principle according to which the transformation laws (3.1)
have to be modified in the special theory of relativity, it is helpful to consider two
coordinate systems in the xy-plane whose axes are rotated relative to each other by
a given angle. They are depicted in Fig. 3.2 together with two points A and B.

The route from point A to point B corresponds to a two-component vector $\vec{r} = \overrightarrow{AB}$:
$\vec{r} = \begin{pmatrix} r_x \\ r_y \end{pmatrix} = \begin{pmatrix} \Delta x^{\text{AB}} \\ \Delta y^{\text{AB}} \end{pmatrix}$. The numerical values of the intervals along the x-axis Δx^{AB}
and the y-axis Δy^{AB} differ in the different coordinate systems (for fixed points A
and B):

Fig. 3.2 The route from A to
B in two different coordinate
systems in the xy-plane

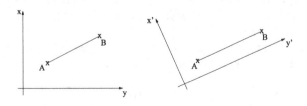

$$\begin{pmatrix} \Delta x^{AB} \\ \Delta y^{AB} \end{pmatrix} \neq \begin{pmatrix} \Delta x^{AB'} \\ \Delta y^{AB'} \end{pmatrix}$$

However, no privileged coordinate system exists in empty space; every "observer" can choose a coordinate system according to his taste, and different observers will measure different quantities Δx^{AB} and Δy^{AB} in general. On the other hand there exists a quantity that is the same in every coordinate system, namely the distance Δ^{AB} between A and B:

$$\left(\Delta^{AB}\right)^2 = |\overrightarrow{AB}|^2 = \left(\Delta x^{AB}\right)^2 + \left(\Delta y^{AB}\right)^2 = \left(\Delta x^{AB'}\right)^2 + \left(\Delta y^{AB'}\right)^2. \quad (3.3)$$

In a coordinate system where $\Delta x^{AB'} = 0$, the expression for the distance simplifies: $\Delta^{AB} = \Delta y^{AB'}$.

Let us briefly consider the effect of a detour: if one person travels directly from A to B while another person takes a detour via C, both have covered the same interval Δy^{AB}, but the total length of the route via C is obviously longer, see Fig. 3.3.

Let us return to the xt-plane. Here one is not obliged to introduce the concept of a "distance" Δ^{AB} that is independent of coordinate systems, but precisely this is done in special relativity. It is denoted as $\Delta \tau^{AB}$, and depends on the spatial distance Δx^{AB} as well as the chronological interval Δt^{AB}. Compared to the formula (3.3) for the distance in the xy-plane, the formula for $\Delta \tau^{AB}$ contains a factor $-1/c^2$, where c is the speed of light:

$$\left(\Delta \tau^{AB}\right)^2 = \left(\Delta t^{AB}\right)^2 - \frac{1}{c^2}\left(\Delta x^{AB}\right)^2, \quad (3.4)$$

where Δt^{AB} and Δx^{AB} are measured in an arbitrary coordinate system, for instance in ours.

We have already seen that spatial distances Δx^{AB} between two events differ in coordinate systems that are moving relative to each other. In special relativity, also chronological intervals Δt^{AB} between the same events but measured in coordinate systems in relative motion are different. However, the same result for $\Delta \tau^{AB}$ is obtained in both coordinate systems if the formula (3.4) for $\Delta \tau^{AB}$ is used. This

Fig. 3.3 A direct route from A to B, and a detour via C

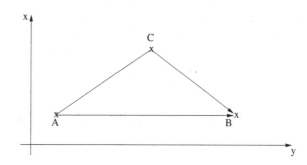

allows us to deduce how the transformation laws (3.1) of Newtonian mechanics have to be modified.

Before we give these modified transformation laws, we will discuss the physical meaning of $\Delta\tau^{AB}$. Recall that, in the xy-plane, the expression for the distance is simply given by $\Delta^{AB} = \Delta y^{AB'}$ in the primed coordinate system in Fig. 3.2, where $\Delta x^{AB'}$ vanishes. A similar coordinate system in the xt-plane is that of the astronaut, in which $\Delta x_{astr}^{AB} = 0$. Only in this coordinate system the formula (3.4) for $\Delta\tau^{AB}$ simplifies to

$$\Delta\tau^{AB} = \Delta t_{astr}^{AB}. \tag{3.5}$$

Correspondingly, $\Delta\tau^{AB}$ is the chronological interval between the events A and B measured by the person (and only by this person) at the site of both events. $\Delta\tau^{AB}$ is also denoted as the *proper time*.

Now we turn to the transformation laws (3.1), which are modified in special relativity and which we rewrite, for simplicity, with fewer indices:

$$\Delta x' = \Delta x - v_x \Delta t, \quad \Delta t' = \Delta t. \tag{3.6}$$

The modified transformation laws in special relativity are denoted as *Lorentz transformations*:

$$\Delta x' = \gamma\left(\Delta x - v_x\Delta t\right), \quad \Delta t' = \gamma\left(\Delta t - \frac{v_x}{c^2}\Delta x\right), \quad \gamma = \frac{1}{\sqrt{1 - \frac{v_x^2}{c^2}}}. \tag{3.7}$$

First it is easy to see that, for velocities v_x small compared to the speed of light c, one finds $\gamma \sim 1$ and the second term in $\Delta t'$ is negligibly small; in this limiting case one re-obtains the relations (3.6).

In addition one can verify that the intervals Δx, Δt and $\Delta x'$, $\Delta t'$, related as in (3.7), lead to the same result for $\Delta\tau$ if $\Delta\tau$ is computed as in (3.4):

$$(\Delta\tau)^2 = (\Delta t)^2 - \frac{1}{c^2}(\Delta x)^2 = \left(\Delta t'\right)^2 - \frac{1}{c^2}\left(\Delta x'\right)^2 = \left(\Delta\tau'\right)^2. \tag{3.8}$$

In this sense $\Delta\tau$ is a distance in the xt-plane independent of the coordinate system.

In (3.7), v_x is the relative velocity between two coordinate systems, neither of which has to coincide with the space ship within which the two events A and B take place. Only if we substitute for v_x the relation (3.2) (and $\Delta x = \Delta x_{us}^{AB}$, $\Delta t = \Delta t_{us}^{AB}$) can we determine from (3.7) the intervals between the events measured by the astronaut in the space ship. For Δx_{astr}^{AB} we still obtain $\Delta x_{astr}^{AB} = 0$ (as must be the case), but the new result for Δt_{astr}^{AB} (using $\Delta x_{us}^{AB} = v_x \Delta t_{us}^{AB}$) is given by

$$\Delta t_{astr}^{AB} = \gamma\left(\Delta t_{us}^{AB} - \frac{v_x}{c^2}\Delta x_{us}^{AB}\right) = \gamma\left(1 - \frac{v_x^2}{c^2}\right)\Delta t_{us}^{AB} = \gamma^{-1}\Delta t_{us}^{AB}. \tag{3.9}$$

Fig. 3.4 Direct route from A
to B, and a detour via C

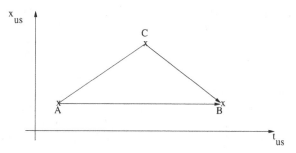

From $\gamma^{-1} < 1$ one obtains

$$\Delta t_{\text{astr}}^{\text{AB}} < \Delta t_{\text{us}}^{\text{AB}}, \tag{3.10}$$

hence the time that passes for the astronaut between the *same* events A and B is
shorter than the chronological time interval measured by us! (The relation between
$\Delta t_{\text{astr}}^{\text{AB}}$ and $\Delta t_{\text{us}}^{\text{AB}}$ is not symmetric, since only the astronaut is at the site of the events
A and B.)

Let us now compare the "travel" of two people from A to B, where A and B
represent two events on Earth with $t_{\text{B}} > t_{\text{A}}$. We simply remain at the same place,
but an astronaut travels via C as indicated in Fig. 3.4.

In the case of Fig. 3.3 in the xy-plane, the total distance via C (the sum of the
distances $\Delta^{\text{AC}} + \Delta^{\text{CB}}$) was longer than the direct route from A to B. Here the
proper time that passes between A and B is not the same but, owing to the inequality
(3.10)—valid for the part AC as well as for the part CB—the proper time for the
astronaut traveling via C is shorter than for the person remaining "on site" on Earth:
$\Delta\tau^{\text{AC}} + \Delta\tau^{\text{CB}} < \Delta\tau^{\text{AB}}$! However, the effect is numerically relevant only if Δt^{AC}
is of the same order of magnitude as $\Delta x^{\text{AC}}/c$, that is, if the velocity of the person
traveling via C is close to the speed of light c.

Finally we return to the Lorentz transformations (3.7) relating two general coordinate systems, for neither of which we assume $\Delta x = 0$ or $\Delta x' = 0$. Let us assume that
the measured intervals in the first coordinate system satisfy the relation $\Delta x/\Delta t = c$.
Hence the measured velocity of the (fictitious) space ship within which the events
took place is equal to the speed of light c. For the proper time that has passed for the
astronaut inside the space ship one obtains $\Delta\tau = 0$.

Now one finds from (3.7)—or, more simply, from (3.8) with $\Delta\tau = 0$—that the
intervals measured in a primed coordinate system satisfy *also* the relation $\Delta x'/\Delta t' = c$: even though the primed coordinate system moves with a velocity v_x relative to
the first coordinate system, the measured velocity of the space ship in the primed
coordinate system is *also* equal to the speed of light! Actually this independence
of the measured speed of light from the coordinate system was at the origin of the
special theory of relativity.

Originally it was assumed that light propagates in a kind of "ether". However, an
ether would define a particular coordinate system in which the ether is at rest. Then

the speed of light measured on Earth would depend on the velocity of the Earth with respect to the ether, and notably on the direction of the light rays. (One concluded from the motion of the Earth around the Sun that the relative velocity of the Earth with respect to the ether cannot always vanish.) On the other hand it was shown in the Michelson–Morley experiment that the speed of light does *not* depend on the direction of the light rays.

Einstein has shown that, in the special theory of relativity, the measured speed of light can always be the same without the hypothesis of an ether if one abandons the idea of absolute time, independent of the coordinate system: in the case of a transformation into a coordinate system moving with a relative velocity v_x according to (3.7), the spatial and chronological axes have to be rotated, similar to possible rotations of x- and y-axes in the xy-plane.

Up to know we have assumed just one spatial dimension x. It is possible to generalize the above equations for the (realistic) case of three dimensions x, y, and z. Now the distances between two events are characterized by Δt, Δx, Δy, and Δz, and (3.4) for the proper time $\Delta \tau$ (still measured by a person on site in a coordinate system where $\Delta x' = \Delta y' = \Delta z' = 0$) becomes

$$\Delta \tau^2 = (\Delta t)^2 - \frac{1}{c^2} \left((\Delta x)^2 + (\Delta y)^2 + (\Delta z)^2 \right). \tag{3.11}$$

The velocity (assumed to be constant) of the person on site, measured in another coordinate system, is a vector:

$$\vec{v} = \begin{pmatrix} v_x \\ v_y \\ v_z \end{pmatrix} = \begin{pmatrix} \Delta x / \Delta t \\ \Delta y / \Delta t \\ \Delta z / \Delta t \end{pmatrix}. \tag{3.12}$$

The modulus of this vector is given by $|\vec{v}|^2 = \vec{v}^2 = v_x^2 + v_y^2 + v_z^2 = \left(\frac{\Delta x}{\Delta t} \right)^2 + \left(\frac{\Delta y}{\Delta t} \right)^2 + \left(\frac{\Delta z}{\Delta t} \right)^2$, which allows us to write (3.11) in the form

$$\Delta \tau^2 = (\Delta t)^2 \left(1 - \frac{\vec{v}^2}{c^2} \right). \tag{3.13}$$

The Lorentz transformations (3.7) can be generalized as well, but the corresponding expressions become quite complicated unless the relative velocity between the two coordinate systems is parallel to one of the axes.

Next we will show that the proper time (3.11) can be interpreted as a distance in a four-dimensional space–time.

First we recall the scalar product of two three-dimensional vectors:

$$\vec{a} = \begin{pmatrix} a_x \\ a_y \\ a_z \end{pmatrix}, \quad \vec{b} = \begin{pmatrix} b_x \\ b_y \\ b_z \end{pmatrix}, \quad \vec{a} \cdot \vec{b} = a_x b_x + a_y b_y + a_z b_z, \tag{3.14}$$

from which we obtain for the modulus of a vector

$$|\vec{a}|^2 = \vec{a} \cdot \vec{a} = a_x^2 + a_y^2 + a_z^2. \qquad (3.15)$$

These formulas can be generalized to a four-dimensional space:

$$\vec{a} = \begin{pmatrix} a_1 \\ a_2 \\ a_3 \\ a_4 \end{pmatrix}, \quad \vec{b} = \begin{pmatrix} b_1 \\ b_2 \\ b_3 \\ b_4 \end{pmatrix}, \quad \vec{a} \cdot \vec{b} = a_1 b_1 + a_2 b_2 + a_3 b_3 + a_4 b_4, \qquad (3.16)$$

from which we obtain for the modulus of a vector

$$|\vec{a}|^2 = \vec{a} \cdot \vec{a} = a_1^2 + a_2^2 + a_3^2 + a_4^2. \qquad (3.17)$$

Now we notice that (3.11) for $\Delta\tau^2$ can be interpreted as the modulus of a four-dimensional vector

$$\Delta\vec{\tau} = \begin{pmatrix} \Delta t \\ \Delta x/c \\ \Delta y/c \\ \Delta z/c \end{pmatrix} \qquad (3.18)$$

under the condition, however, that the rule (3.16) for the scalar product of two four-dimensional vectors is changed as follows:

$$\vec{a} \cdot \vec{b} = a_1 b_1 - a_2 b_2 - a_3 b_3 - a_4 b_4 , \quad \vec{a}^2 = \vec{a} \cdot \vec{a} = a_1^2 - a_2^2 - a_3^2 - a_4^2. \qquad (3.19)$$

Now we find

$$\Delta\vec{\tau}^2 = \Delta\vec{\tau} \cdot \Delta\vec{\tau} = (\Delta t)^2 - \left(\frac{\Delta x}{c}\right)^2 - \left(\frac{\Delta y}{c}\right)^2 - \left(\frac{\Delta z}{c}\right)^2, \qquad (3.20)$$

in agreement with (3.11).

A space in which scalar products are computed according to the formula (3.19) is denoted as *Minkowski space*.

Nevertheless one can

(a) compute sums and differences between two vectors as usual,
(b) perform coordinate transformations as rotations of axes that leave lengths (moduli) of vectors invariant. If the time-like axis is rotated into a space-like axis in this process, the coordinate transformations correspond to the Lorentz transformations.

Accordingly it is possible, in the framework of special relativity, to interpret space–time as a four-dimensional Minkowski space. However, a Minkowski space cannot be represented graphically—even in two dimensions. For this reason it is hopeless to deduce the above formulas from drawings: every sheet of paper is always an Euclidean space in which the equations (3.3) for distances are valid—that is why the reduced proper time for the trajectory via C in Fig. 3.4 cannot be understood graphically.

An additional peculiarity of Minkowski space is that the modulus of a vector is not necessarily positive. In the case of the proper time (3.20), which represents a measurable quantity in a given coordinate system, we require, however, that $\Delta\tau^2$ is (semi)positive such that the root can be extracted and that $\Delta\tau$ can be computed. It follows from (3.13) that $\Delta\tau^2$ is semi-positive only if $|\vec{v}| \leq c$. If an object were to move faster than the speed of light, it would become impossible to obtain a reasonable result for $\Delta\tau$.

3.1.1 Energy and Momentum

Given an object with mass m and velocity $\vec{v}$, the following quantities are defined in classical mechanics:

(a) the kinetic energy $E_{\text{kin}} = \frac{1}{2}m\vec{v}^2$,
(b) the momentum $\vec{p} = m\vec{v}$.

Correspondingly the kinetic energy can be expressed directly in terms of the momentum:

$$E_{\text{kin}} = \frac{1}{2m}\vec{p}^2. \tag{3.21}$$

This relation is no longer valid in special relativity. The energy originating from the mass has to be added, and the dependence of the energy on the momentum $\vec{p}$ is given by

$$E^2 = m^2c^4 + \vec{p}^2c^2 \quad \text{or} \quad E = \sqrt{m^2c^4 + \vec{p}^2c^2}. \tag{3.22}$$

If the velocity is much smaller than the speed of light c, we have $\vec{p}^2 \ll m^2c^2$, and the expression for E can be expanded in a series in $\vec{p}^2/m^2c^2$:

$$E = mc^2\sqrt{1 + \frac{\vec{p}^2}{m^2c^2}} \simeq mc^2\left(1 + \frac{\vec{p}^2}{2m^2c^2} + \dots\right) = mc^2 + \frac{\vec{p}^2}{2m} + \dots \tag{3.23}$$

In this equation we find the contribution of the mass to the energy ($E = mc^2$ for $\vec{p} = 0$), as well as the "classical" expression for the kinetic energy in the second term: hence the classical expression is not wrong in special relativity, but applicable "only" to the case of velocities small compared to the speed of light. (The first momentum-independent term mc^2 plays no role for energy differences at constant mass, and only energy differences count in mechanics.)

It is remarkable that the first expression for E in (3.22) can be written in the following form:

$$m^2c^2 = \frac{1}{c^2}E^2 - \vec{p}^2 = \vec{P_4}^2, \tag{3.24}$$

where the components of the "energy momentum vector" $\vec{P_4}$ are given by

$$\vec{P_4} = \begin{pmatrix} \frac{1}{c}E \\ p_x \\ p_y \\ p_z \end{pmatrix}, \tag{3.25}$$

and $\vec{P_4}^2$ has to be computed according to the rule (3.19) for Minkowski space. Thus the modulus of the energy momentum four vector is constant, and given by m^2c^2.

Finally we note that the relation $\vec{p} = m\vec{v}$ between momentum and velocity is no longer valid in special relativity, but has to be replaced by

$$\vec{p} = m\vec{v} / \sqrt{1 - \frac{\vec{v}^2}{c^2}}, \tag{3.26}$$

which can be written as

$$\vec{v}^2 = \frac{c^2\vec{p}^2}{(m^2c^2 + \vec{p}^2)}. \tag{3.27}$$

In particular, a massless particle like the photon (whose velocity is always equal to c) possesses a non-vanishing momentum related to its energy according to (3.22) by

$$E = |\vec{p}|\,c \quad (\text{if } m = 0). \tag{3.28}$$

Even in the case of a massless particle, energy and momentum are still variable!

What happens in the case of a massive particle if its energy is very large (much larger than mc^2)? It follows from the relation (3.22) between energy and momentum $\vec{p}$ that the modulus $|\vec{p}|$ of the momentum is very large as well, and related to the energy approximately as in Eq. (3.28) valid for a massless particle. However, even for arbitrarily large momentum $|\vec{p}|$, the modulus $|\vec{v}|$ of the velocity never exceeds the speed of light c, as we can see from (3.27) in the limit $\vec{p}^2 \gg m^2c^2$: in special relativity the velocity of a massive object (or a massive particle) is always smaller than the speed of light, and the velocity of a massless object (or a massless particle) always equal to the speed of light.

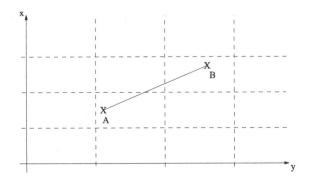

Fig. 3.5 The distance between A and B in a Cartesian coordinate system

Fig. 3.6 The surface of a sphere does *not* allow for a Cartesian coordinate system

3.2 The General Theory of Relativity: Curved Spaces

The formula for the distance between two points A and B in a surface,

$$\left(\Delta^{AB}\right)^2 = \left(\Delta x^{AB}\right)^2 + \left(\Delta y^{AB}\right)^2, \tag{3.29}$$

is only valid if we employ a Cartesian coordinate system. In a Cartesian coordinate system the angle between any two lines parallel to the coordinate axes x and y, respectively, is 90° *everywhere* in the surface (see Fig. 3.5).

This is impossible in a curved surface such as the surface of a sphere in Fig. 3.6.

In a curved surface the sum of the angles inside a triangle differs from 180°, and we talk about a non-Euclidean or Riemannian geometry. Attention: a rolled up sheet of paper is still *flat* in this sense; the deformation of a sheet into a curved surface always generates cracks or wrinkles.

In the case of a curved surface, the formula (3.29) for the distance Δ^{AB} in terms of Δx^{AB} and Δy^{AB} is no longer valid. Now the distance Δ^{AB} depends in a more complicated way on Δx^{AB} and Δy^{AB}:

$$\left(\Delta^{AB}\right)^2 = g_{xx}\left(\Delta x^{AB}\right)^2 + 2g_{xy}\left(\Delta x^{AB}\right)\left(\Delta y^{AB}\right) + g_{yy}\left(\Delta y^{AB}\right)^2 \tag{3.30}$$

Δx^{AB} and Δy^{AB} are intervals along the coordinate axes (which depend on the coordinate system) between the points A and B. However, the distance Δ^{AB} is the same in every coordinate system by definition.

The three coefficients g_{xx}, g_{xy}, and g_{yy} are functions of the coordinates of the points A and B in general. Hence one prefers to consider (3.30) in the limit where the points A and B are very close, all "Δ" in (3.30) tend to zero, but their ratios remain finite. In this limit the three coefficients g_{xx}, g_{xy}, and g_{yy} are functions of the coordinates of the points A or B; the difference becomes negligible. The corresponding functions are denoted as the *metric* of the surface. (The metric also depends on the coordinate system, such that the distance Δ^{AB} is independent of the coordinate system.)

Given the metric (i.e., the three functions in the case of a two-dimensional surface), the curvature and all other properties of the surface can be computed. The surface is flat, or curvature-free, if there exists a coordinate system where $g_{xx} = g_{yy} = 1$ and $g_{xy} = 0$ holds. In this case (3.30) for Δ^{AB} coincides with (3.29).

(Even if the surface is curvature-free, the metric is not necessarily of this simple form in a non-Cartesian coordinate system: in polar coordinates r and θ, $\left(\Delta^{AB}\right)^2$ in a flat plane is given by $\left(\Delta r^{AB}\right)^2 + r^2 \left(\Delta \theta^{AB}\right)^2$. Hence we have $g_{rr} = 1$ and $g_{r\theta} = 0$, but $g_{\theta\theta} = r^2$.)

It is possible to generalize this formalism to three- and four-dimensional spaces. In the four-dimensional case we denote the coordinates by 0, 1, 2, 3, where the coordinate 0 corresponds to the time t. The expression for the distance Δ^{AB}, generalizing (3.30), becomes

$$\left(\Delta^{AB}\right)^2 = g_{00}\left(\Delta_0^{AB}\right)^2 + 2g_{01}\Delta_0^{AB}\Delta_1^{AB} + \ldots + g_{22}\left(\Delta_2^{AB}\right)^2$$
$$+ 2g_{23}\Delta_2^{AB}\Delta_3^{AB} + \ldots \tag{3.31}$$

Here Δ_0^{AB}, Δ_1^{AB}, etc. denote the intervals along the axes 0, 1, etc. Obviously it becomes tedious to list all terms explicitly; therefore we employ the compact notation

$$\left(\Delta^{AB}\right)^2 = \sum_{\mu=0}^{3}\sum_{\nu=0}^{3} g_{\mu\nu}\,\Delta_\mu^{AB}\,\Delta_\nu^{AB}. \tag{3.32}$$

(Usually Greek indices $\mu, \nu, \ldots$ can assume four different values, whereas Latin indices $i, j, \ldots$ are assigned to coordinates of three-dimensional spaces.) A priori 16 terms appear on the right-hand side of this equation; one can assume, however, that the metric $g_{\mu\nu}$ is symmetric (that $g_{\mu\nu} = g_{\nu\mu}$ holds), whereupon the metric contains "only" 10 independent functions in four dimensions.

Let us recall the formula (3.11) for the proper time (the invariant distance) between two events A and B, expressed as a length of the four-vector $\Delta\vec{\tau}$ with the help of the modified rule (3.19) in (3.20). We can identify the invariant distance as Δ^{AB} in (3.32), and write the modified rule (3.19) in terms of correspondingly chosen functions $g_{\mu\nu}$ (which are constants here):

$$g_{00} = 1, \quad g_{11} = g_{22} = g_{33} = -1, \quad g_{\mu\nu} = 0 \quad \text{if } \mu \neq \nu. \tag{3.33}$$

This metric is denoted as the *Minkowski metric* and describes a flat space–time.

In general relativity, $g_{\mu\nu}$ are true functions of the components of the position vector $\vec{r}$ and the time t in general and describe, correspondingly, a curved four-dimensional space–time (or a curved four-dimensional space). The functions $g_{\mu\nu}$ are determined by the Einstein equations; these are equations for $g_{\mu\nu}$ depending on the matter and the energy distributed in space. In the case of a flat homogeneous space with a time-dependent scale factor $a(t)$ one can choose $g_{00} = 1$, $g_{11} = g_{22} = g_{33} = -a^2(t)$, and the Einstein equations result in the Friedmann–Robertson–Walker equations (2.6) and (2.7). However, in the region around a star of mass M the metric is no longer homogeneous, but depends on the distance r to the center of the star. From the Einstein equations one obtains for the component g_{00} (outside the star)

$$g_{00}(r) = 1 - \frac{2GM}{rc^2}, \tag{3.34}$$

where G is Newton's constant:

$$G \simeq 6.67 \times 10^{-11} \, \mathrm{m}^3 \, \mathrm{kg}^{-1} \, \mathrm{s}^{-2}. \tag{3.35}$$

Now the formula (3.20) for the proper time between two events A and B has to be replaced by

$$\Delta\tau^2 = g_{00}(r)\,(\Delta t)^2 + g_{xx}(r)\left(\frac{\Delta x}{c}\right)^2 + g_{yy}(r)\left(\frac{\Delta y}{c}\right)^2 + g_{zz}(r)\left(\frac{\Delta z}{c}\right)^2, \tag{3.36}$$

where (3.34) has to be used for g_{00}. (In the coordinate system used here, the deviations of the components g_{xx}, g_{yy}, and g_{zz} from -1 play no role in the following as long as we restrict ourselves to velocities small compared to the speed of light.)

We recall the meaning of the terms in (3.36): the proper time $\Delta\tau$ is the time that passes between two events A and B as measured by an astronaut in the space ship in which the events took place. For this astronaut the spatial distance between the events vanishes. Δt, Δx, etc. are the chronological and spatial intervals, respectively, between the same events as measured by an observer with respect to whom the space ship moves with a given velocity $v_x = \Delta x / \Delta t$.

Since, according to (3.36), the relation between Δt and $\Delta\tau$ depends on the position r, the observer has the impression that the velocity of the space ship does not remain constant. One can show that the observer observes an *acceleration* of the space ship, and that the components of the vector $\vec{a}$ of the observed acceleration are given by

$$a_x = -\frac{c^2}{2}\frac{\partial}{\partial x}g_{00}(r), \quad a_y = -\frac{c^2}{2}\frac{\partial}{\partial y}g_{00}(r), \quad a_z = -\frac{c^2}{2}\frac{\partial}{\partial z}g_{00}(r). \tag{3.37}$$

It is possible to combine these three equations with help of a vector $\vec{\nabla}$ defined as

$$\vec{\nabla} = \begin{pmatrix} \frac{\partial}{\partial x} \\ \frac{\partial}{\partial y} \\ \frac{\partial}{\partial z} \end{pmatrix} . \tag{3.38}$$

Then (3.37) simplifies to

$$\vec{a} = -\frac{c^2}{2}\vec{\nabla} g_{00}(r) . \tag{3.39}$$

Accordingly the observer has the impression that a force

$$\vec{F} = m\vec{a} = -\frac{mc^2}{2}\vec{\nabla} g_{00}(r) = -\frac{GmM\vec{r}}{r^3} \tag{3.40}$$

acts on the space ship, where m is the mass of the space ship.

This force is nothing but gravity!

However, the astronaut in his otherwise propulsionless space ship feels neither an acceleration nor a force; initially the above force is just an interpretation of the observer. We feel the gravitational force on the surface of the Earth if and only if a resistance, for example the floor, acts against the natural motion, free fall.

Since gravity originates from the curvature of space–time as described by the metric (3.34), *every* object is subject to the gravitational force: even massless particles such as photons feel the curvature of space–time near stars and planets. One can verify the bending of light rays that reach us from a distant star along a trajectory that passes close to the Sun. The measured bending angle is in agreement with general relativity, but not with Newtonian mechanics, which would predict only about half the angle.

Since the gravitational acceleration is independent of the mass, composition, and inner structure of a body, it fulfills the so-called equivalence principle (which corresponds to this statement), and which has been verified to very high precision.

3.2.1 Black Holes

Let us assume that we are very far from a star, and that we drop an object with negligibly small initial velocity in the direction of the star. Subsequently we want to determine the radial component of the velocity v of the object as a function of the distance r to the center of the star. To this end we write (a is the acceleration dv/dt)

$$\frac{dv}{dr} = \frac{dv}{dt}\frac{dt}{dr} = av^{-1}, \quad \text{implying} \quad v\frac{dv}{dr} = a \quad \text{or} \quad \frac{1}{2}\frac{d}{dr}v^2 = a. \tag{3.41}$$

Following (3.39), the acceleration a is given by $-\frac{c^2}{2}\frac{d}{dr}g_{00}(r)$, and we obtain

$$\frac{d}{dr}v^2 = -c^2\frac{d}{dr}g_{00}(r), \quad \text{hence} \quad v^2 = -c^2 g_{00}(r) + C \quad \text{or}$$

$$v^2 = -c^2 + \frac{2GM}{r} + C, \tag{3.42}$$

where we have used (3.34) for $g_{00}(r)$. The constant C is determined by the initial condition, a negligibly small initial velocity $v = 0$ at a very large distance $r \to \infty$ ($C = c^2$), and we obtain

$$v^2 = \frac{2GM}{r}. \tag{3.43}$$

If r becomes smaller and smaller, the object either crashes onto the surface of the star or—if the radius of the star is sufficiently small—the velocity of the object increases more and more and seems to exceed the speed of light! According to (3.43) this happens at a distance denoted as the *Schwarzschild radius* r_S,

$$r_S = \frac{2GM}{c^2}, \tag{3.44}$$

of the object to the center of the star.

In reality we can never observe an object with a velocity exceeding the speed of light c. Here the following happens: beginning with the moment when its velocity becomes seemingly larger than c, the light emitted by the object can no longer reach us—the object becomes invisible, i.e., "black". Also, from this moment onwards the object can never return to us!

A star with a radius smaller than the Schwarzschild radius, the condition for this phenomenon to take place, is denoted as a *black hole*. In addition (it does not follow directly from the consideration above) we can show that the light (the photons) emitted from the surface of the star cannot reach a distance from the star larger than the Schwarzschild radius, thus the complete star seems invisible. However, matter (dust and stars) falling into the black hole can heat up before reaching the Schwarzschild radius by the influence of the gravitational forces. As a result of the resulting accelerations and collisions, radiation is emitted that can be used to detect black holes (in addition to the measurable acceleration of nearby stars, for example in the center of our Milky Way).

Our Sun, which has a mass of about 2×10^{30} kg, would be a black hole only if its entire mass were concentrated within a sphere of radius about 3 km (instead of its actual radius of about 7×10^5 km). Similarly, in the case of the Earth, its mass of about 6×10^{24} kg would have to be concentrated in a sphere of radius about 9 mm.

We should add that the above derivation of the Schwarzschild radius is somewhat heuristic; in special relativity, the relation between energy and velocity differs somewhat from the form assumed above, and the expression for the complete Schwarzschild metric is somewhat more complicated. Nevertheless the result (3.44) for the Schwarzschild radius is—accidentally—exact.

We should underline that the existence of black holes is not an obscure speculation within the framework of general relativity but an automatic consequence of the metric

(3.34) in the environment of a star, which is, incidentally, also responsible for the "ordinary" gravitational force.

Exercises

3.1 Verify the assertion in (3.8) that the expression for the proper time $\Delta\tau$ is invariant under a Lorentz transformation (3.10).

3.2 Find the Schwarzschild radius of an object of a mass of 1 kg. (If you compress a mass of 1 kg inside a sphere of this radius, you create a black hole.) Compare r_S to the radius of an atom and the radius of a nucleus, respectively.

Chapter 4
The Theory of Fields

Fields play a fundamental role in the modern formulation of fundamental interactions. We introduce the basic equation of motion for fields, and discuss their most important solutions: the wave solution is relevant both for electromagnetic radiation and for the description of a beam of particles within the framework of quantum field theory. The Coulomb solution describes fields around point-like bodies. The existence of a wave solution for the gravitational field leads to the prediction of the existence of gravitational waves. These are being searched for in experiments being carried out today; the design of these experiments is sketched.

4.1 The Klein–Gordon Equation

A quantity defined at any position $\vec{r}$ (i.e., for all x, y, and z) and at any time t, i.e., which assumes a certain value at any position $\vec{r}$ at any time t, is denoted as a "field". It is written, e.g., as $\Phi(\vec{r}, t)$. Known fields are the temperature $T(\vec{r}, t)$, the pressure $p(\vec{r}, t)$, the velocity of the wind $\vec{v}(\vec{r}, t)$, and the electric field $\vec{E}(\vec{r}, t)$. In the latter two cases the field is oriented along a certain direction, which depends, in general, on the position and on time. Such fields are denoted as *vector fields*, whereas $T(\vec{r}, t)$ and $p(\vec{r}, t)$ are denoted as *scalar fields*.

Here the fields $T(\vec{r}, t)$, $p(\vec{r}, t)$, and $\vec{v}(\vec{r}, t)$ are not fundamental fields, but simplified descriptions (averages) of complicated motions of a countless number of atoms or molecules. However, the electric field $\vec{E}(\vec{r}, t)$ is a fundamental field in the sense that it generates one of the fundamental forces of nature (see Chap. 5).

Such a field is usually invisible; however, it can be detected with the help of objects on which it exerts a force. The presence of an electric field is measurable only with the help of charged objects, which allow the force exerted by the field to be measured.

The components $g_{\mu\nu}(\vec{r}, t)$ of the metric, discussed in the previous chapter on general relativity, are fields as well. These fields play two roles: on the one hand they determine the curvature of space-time, and on the other hand the field $g_{00}(\vec{r}, t)$ is

U. Ellwanger, *From the Universe to the Elementary Particles*, 45
Undergraduate Lecture Notes in Physics, DOI: 10.1007/978-3-642-24375-2_4,
© Springer-Verlag Berlin Heidelberg 2012

related via (3.40) to the gravitational force. Since the curvature of space-time affects the trajectory of any object (even of objects of vanishing mass), any object can serve to detect a "gravitational field", i.e., a deviation of the metric from the flat Minkowski metric (3.33).

In general every field satisfies a differential equation containing derivatives with respect to the time t and the coordinates x, y, and z. This equation determines the time evolution (the second derivative with respect to t) at any position, depending on its value and its spacial derivatives in this point. (For pointlike objects, the second derivative with respect to time of the position vector, the acceleration, is given by the force acting on the object and the equation $\vec{a} = \frac{1}{m}\vec{F}$.)

The fundamental fields (denoted by $\Phi(\vec{r}, t)$ in the following) satisfy—under certain conditions, such as the absence of a mass term and neglect of couplings to other fields—a differential equation denoted as the *Klein–Gordon equation*:

$$\left(\frac{\partial^2}{\partial t^2} - c^2\left(\frac{\partial^2}{\partial x^2} + \frac{\partial^2}{\partial y^2} + \frac{\partial^2}{\partial z^2}\right)\right)\Phi(\vec{r}, t) = 0, \tag{4.1}$$

where c denotes the speed of light. (Here and below we use partial derivatives in order to emphasize that the variables t, x, y, and z have to be considered as independent.) We can observe a certain similarity of this equation to the right-hand side of (3.11) for the proper time $\Delta\tau$ in special relativity. In fact the Klein–Gordon equation is invariant under coordinate transformations or redefinitions of the variables t, x, y, and z in the form of Lorentz transformations (3.7), which, according to (3.8), also leave the proper time $\Delta\tau$ unchanged.

Below we will discuss two different solutions of this equation.

4.2 The Wave Solution

In order to simplify the equation—without losing the essential properties of this solution—we can assume that $\Phi(\vec{r}, t)$ depends on only one (e.g., on x) of the three spatial coordinates x, y, and z. Then we can replace $\Phi(\vec{r}, t)$ by $\Phi(x, t)$ and use $\frac{\partial}{\partial y}\Phi(x, t) = \frac{\partial}{\partial z}\Phi(x, t) = 0$. The Klein–Gordon equation simplifies to

$$\left(\frac{\partial^2}{\partial t^2} - c^2\frac{\partial^2}{\partial x^2}\right)\Phi(x, t) = 0. \tag{4.2}$$

A solution of this equation is given by

$$\Phi(x, t) = \Phi_0 \cos(\omega t - kx), \tag{4.3}$$

where the constant Φ_0 is called the *amplitude* of the wave, and ω and k have to satisfy a relation we will derive: using $\frac{\partial}{\partial t}\cos(\omega t - kx) = -\omega\sin(\omega t - kx)$, $\frac{\partial}{\partial t}\sin(\omega t - kx) = \omega\cos(\omega t - kx)$, $\frac{\partial}{\partial x}\cos(\omega t - kx) = k\sin(\omega t - kx)$, and $\frac{\partial}{\partial x}\sin(\omega t - kx) = -k\cos(\omega t - kx)$ we obtain from (4.2) with (4.3) for $\Phi(x, t)$

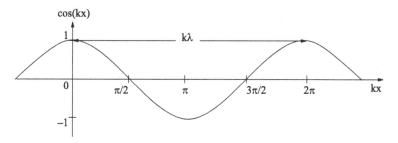

Fig. 4.1 x dependence of a wave at a fixed time t

$$\left(-\omega^2 + c^2 k^2\right) \Phi(x, t) = 0. \tag{4.4}$$

This equation is satisfied for all x and t only if

$$\omega = \pm ck. \tag{4.5}$$

(k or ω can be chosen arbitrarily.) The following considerations are helpful for the physical interpretation of the wave solution (4.3).

(a) First we consider $\Phi(x, t)$ at a fixed time $t = 0$ (corresponding to a "picture" of $\Phi(x, t = 0)$). Then the dependence on x has the form of a wave $\Phi_0 \cos(-kx) = \Phi_0 \cos(kx)$, for which the wavelength λ (the distance between two neighboring maxima or two neighboring minima) is given by

$$\lambda = \frac{2\pi}{k}, \tag{4.6}$$

see Fig. 4.1. The amplitude of the wave is Φ_0.

(b) Let us consider $\Phi(x, t)$ at a fixed position $x = 0$. The dependence on time is given by an oscillation $\Phi_0 \cos(\omega t)$ with period $T = 2\pi/\omega$ (see Fig. 4.2). $1/T$ is equal to the frequency ν, corresponding to the number of oscillations per second. From $T = 1/\nu$ one obtains $\omega = 2\pi \nu$.

(c) Let us consider the motion of a maximum of the wave: a maximum (the first one) of the function $\cos(\varphi)$ is at $\varphi = 0$. Here we have to replace φ by $\omega t - kx$. At the time $t = 0$ the position of the first maximum x_{max} is given by $x_{max} = 0$, since we have $\varphi(x_{max}, t = 0) = 0$. Once t has increased by Δt, the new position Δx of the maximum has to be determined from the condition $\varphi(\Delta x, \Delta t) = \omega \Delta t - k \Delta x = 0$, which gives $\Delta x = \frac{\omega}{k} \Delta t$. Identifying the velocity v of the wave with the velocity of the maximum, we obtain $v = \frac{\Delta x}{\Delta t} = \frac{\omega}{k}$. Using the relation (4.5) between ω and k we find $|v| = c$, hence $|v|$ is equal to the speed of light!

From (4.5), (4.6) and $\omega = 2\pi \nu$ we also obtain a relation between the wavelength λ and the frequency ν:

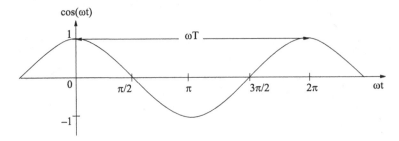

Fig. 4.2 t dependence of a wave at a fixed position x

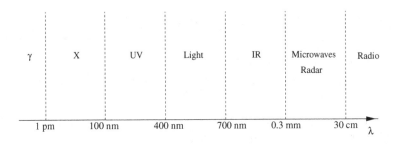

Fig. 4.3 The usual notation for electromagnetic radiation as a function of the wavelength λ. (The division of the axis is *not* to scale)

$$\nu\lambda = c \qquad (4.7)$$

Below we will see that the electric field $\vec{E}(\vec{r}, t)$ and the magnetic field $\vec{B}(\vec{r}, t)$ satisfy the Klein–Gordon equation. Electromagnetic radiation (light, radio waves, ...) corresponds to wave solutions of this equation with different wavelengths λ. The usual notation for electromagnetic radiation is shown in Fig. 4.3 as a function of the wavelength λ.

In Fig. 4.3 we have used the following abbreviations for length units: 1 nm = 1 nanometer = 10^{-9} m, and 1 pm = 1 picometer = 10^{-12} m. The abbreviations for the different kinds of radiation are as follows (from left to right):

γ: gamma radiation
X: X radiation
UV: ultraviolet radiation
Light denotes the various colors of visible light, where 400 nm corresponds to blue light and 700 nm to red light
IR: infrared radiation
Microwave radiation has about the same range of wavelengths as radar waves
Radio indicates the usual range of radio frequencies

With the help of the relation (4.7) it is easy to determine the frequency ν from the wavelength λ for any kind of radiation.

In *quantum field theory* we can identify waves of a field with a beam of corresponding particles. The kinetic energy of the particles depends on ω or on the frequency ν of the wave:

$$E = \hbar\omega = h\nu, \tag{4.8}$$

where $h = 2\pi\hbar$ is *Planck's constant* (after Max Planck, Nobel prize in 1918):

$$h \simeq 6.626 \times 10^{-34}\,\text{kg}\,\text{m}^2\,\text{s}^{-1},\, \hbar \simeq 1.055 \times 10^{-34}\,\text{kg}\,\text{m}^2\,\text{s}^{-1}. \tag{4.9}$$

The momentum of these particles (directed along the x-axis in our case) depends on the wavelength λ:

$$p_x = \hbar k = h/\lambda \tag{4.10}$$

(This expression for the wavelength λ is also denoted as the *de Broglie wavelength*, after L.-V. de Broglie, Nobel prize 1929.)

Thus (4.5) (or (4.7)) implies a relation between the energy E and p_x of the form $E = cp_x$. Since p_x is the only component of the vector $\vec{p}$ here, this relation is equivalent to $E = c\,|\vec{p}|$ (see (3.28)), which is indeed satisfied for a massless particle in special relativity.

It is possible to generalize these relations to particles with mass. Then, the corresponding field satisfies the massive Klein–Gordon equation

$$\left(\frac{\partial^2}{\partial t^2} - c^2\left(\frac{\partial^2}{\partial x^2} + \frac{\partial^2}{\partial y^2} + \frac{\partial^2}{\partial z^2}\right) + \frac{m^2c^4}{\hbar^2}\right)\Phi(\vec{r}, t) = 0, \tag{4.11}$$

which has also a wave solution of the form (4.3) where, however, $\omega^2 = c^2k^2 + m^2c^4/\hbar^2$ holds instead of (4.5). Since (4.8) and (4.10) remain unchanged, the relation between energy and momentum becomes $E^2 = m^2c^4 + \vec{p}^2c^2$ in agreement with (3.22).

It is important that, according to (4.10), the wavelength λ decreases with increasing momentum (and hence with increasing energy). For the following reason small wavelengths are desired for experiments in particle physics: it is known from the optics of microscopes that light of a given wavelength λ can only resolve objects larger than $\sim\lambda/2\pi$. This statement holds also in the case where a beam of particles is directed on objects (fundamental or composite particles): the more precisely we want to study the object, the smaller the wavelength corresponding to the beam of particles, i.e., the larger the required energy of the particles in the beam must be. This explains why we are interested in the largest possible energies E in particle physics, since one obtains for the resolving power Δ

$$\Delta \sim \frac{\lambda}{2\pi} = \frac{\hbar c}{E}. \tag{4.12}$$

4.3 The Coulomb Solution

Let us return to the original equation (4.1). Now we look for a *static* solution, which is independent of the time t. In this case we can replace $\Phi(\vec{r}, t)$ by $\Phi(\vec{r})$ with $\frac{\partial}{\partial t}\Phi(\vec{r}) = 0$. After division by c^2 and a change of sign, the Klein–Gordon equation becomes

$$\left(\frac{\partial^2}{\partial x^2} + \frac{\partial^2}{\partial y^2} + \frac{\partial^2}{\partial z^2}\right)\Phi(\vec{r}) = 0. \tag{4.13}$$

We look for a solution of the form $\Phi(\vec{r}) = \Phi(|\vec{r}|)$ with spherical symmetry around the origin, where $|\vec{r}| = r = \sqrt{x^2 + y^2 + z^2}$. Using $\partial r/\partial x = x/r$ etc., we can verify that the following expression solves (4.13):

$$\Phi(r) = \frac{\Phi_0}{r} + C = \frac{\Phi_0}{\sqrt{x^2 + y^2 + z^2}} + C, \tag{4.14}$$

where Φ_0 and C are two arbitrary constants. What are the properties of this solution? First it is, by construction, symmetric around the origin and depends only on the distance r to the origin. In the limit $r \to \infty$ it approaches the constant C (often chosen as zero). If r approaches zero, the solution is singular, i.e., it approaches infinity.

In the case of an electric point charge, the Coulomb solution allows one to determine the electric field in the surroundings of the charge; this is the origin of the name Coulomb. The precise relation between this solution and the electric field will be discussed in Chap. 5.

4.4 Gravitational Waves

We have already become familiar with the solution (4.14) in the context of general relativity: in the neighborhood of an object of mass M at the origin of the coordinate system, the component $g_{00}(\vec{r})$ of the metric (see (3.34)) is given by

$$g_{00}(\vec{r}) = 1 - \frac{2GM}{c^2 r}, \tag{4.15}$$

which corresponds to the solution (4.14) with $C = 1$, $\Phi_0 = -2GM/c^2$.

Indeed, assuming $|g_{00} - 1| \ll 1$, i.e., $r \gg 2GM/c^2$, the Einstein equations for the components of the metric assume the form of the Klein–Gordon equation.

It follows that wave solutions for the metric should also be possible. These wave solutions are denoted as *gravitational waves*. However, gravitational waves are very difficult to generate: owing to the smallness of the gravitational constant, enormous masses have to be strongly accelerated. Such processes take place only during explosions or collisions of stars, possibly also at the beginning of the Universe during the

Fig. 4.4 Variations of distances in the plane perpendicular to the direction of propagation of a gravitational wave

Big Bang. It is very difficult, however, to predict the amplitudes of the gravitational waves generated under these circumstances.

An exception are systems of two very massive and compact stars (so-called neutron stars) rotating around each other. Here the emission of gravitational waves is calculable, implying a calculable loss of energy. This energy loss generates a small change of the angular velocity (speed of revolution), which is measurable if one of the neutron stars is a so-called pulsar: a pulsar emits electromagnetic radiation in the range of radio frequencies at very regular intervals. The intervals measured on Earth vary slightly due to the revolution of the pulsar around its partner, depending on whether the pulsar is moving towards us or away from us. This allows a very precise measurement of its revolution period (of the order of several hours). This revolution period should change by about 10^{-4} s per year due to the emission of gravitational waves, and the measured variations agree well with the calculations. (R.A. Hulse and J.H. Taylor received the Nobel prize in 1993 for the confirmation of this phenomenon.) Up to now this is the only—though very indirect—verification of gravitational waves.

The direct detection of gravitational waves is not easy. A gravitational wave passing between two objects generates a small time-dependent variation of the measured distance between the objects in a plane perpendicular to its propagation direction. The frequency of this variation of the distance is equal to the frequency of the gravitational wave, which is related to its wavelength as in (4.7).

Gravitational waves have the following particular property: whenever a distance along a given direction is reduced for a short time, the distance along the perpendicular direction (in the plane perpendicular to the direction of propagation) increases. This behavior is sketched in Fig. 4.4, where we have to imagine all circles and ellipses in the plane perpendicular to the direction of propagation.

At present, experiments for the detection of gravitational waves are carried out using laser beams. In order to understand the mode of operation of such experiments, it is necessary to know about the particular properties of laser beams.

Ordinary light—even light of a given color—consists of a mixture of various waves whose peaks and troughs are distributed erratically along the beam axis. Laser beams are light of a given wavelength resp. color (there exist laser beams of different colors) consisting of a single wave with corresponding wave peaks and troughs.

If we superimpose two different laser beams of the same wavelength with the help of semipermeable mirrors (i.e., if we combine them into a single beam), the result is very sensitive to the relative positions of the wave peaks and troughs of the original beams along the now common beam axes: if the positions of the wave

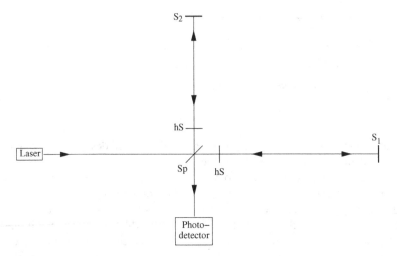

Fig. 4.5 Sketch of a design of a gravitational wave detection experiment

peaks and troughs coincide, the beams enforce each other; if these positions do not coincide, the intensity of the common beam gets reduced. In the extreme case where the peaks of one beam encounter the troughs of the other beam (and hence vice versa), the intensity of the common beam vanishes, i.e., the two waves delete each other. This phenomenon is known as interference. It is possible to obtain a situation of complete deletion from a situation of maximal amplification by shifting the original beams by half a wavelength relative to each other—which allows extremely precise measurements of relative distances.

The design of a gravitational wave detection experiment is sketched in Fig. 4.5 in a view from above.

In Fig. 4.5 a single laser beam is generated on the left. This beam falls onto a semipermeable mirror Sp (the beam splitter), which splits the beam into one traveling to the right and one traveling upwards. At first we can neglect the semipermeable mirrors hS. Next the beams encounter the fully reflecting mirrors S_1 and S_2, are reflected, and meet again at Sp. With the help of the semipermeable mirror Sp they can rejoin and form a beam traveling downwards, whose intensity is measured by the photodetector.

Whenever a gravitational wave passes through this apparatus, the distance Sp–S_1 will be reduced for a short time and the distance Sp–S_2 will be increased for a short time, and vice versa due to the properties of the gravitational wave sketched in Fig. 4.4. In both cases, the distances covered by the two beams have changed relative to the other one. Due to the interference phenomenon discussed above, a relative variation of the distances covered by the two beams of half a laser wavelength (in the range of wavelengths of visible light) suffices to generate measurable variations of the intensity in the photodetector of the unified beam.

The semipermeable mirrors hS reflect the beams again after they have been reflected by S_1 and S_2, such that the distances hS–S_1 and hS–S_2 are covered up to 1000 times before the beams join again at Sp. In this way, the difference of the distances covered by the beams, generated by a gravitational wave, is amplified by a factor of 1000.

At present, several such gravitational wave detectors are running worldwide: GEO600 near Hanover in Germany (with arm lengths Sp–S_1 = Sp–S_2 of 600 m), two identical detectors LIGO Hanford and LIGO Livingston with arm lengths of 4 km in the USA, TAMA with an arm length of 300 m in Japan, and Virgo with an arm length of 3 km near Cascina in Italy. The internet addresses of these experiments, where more precise information can be found, are given in the appendix.

It absolutely necessary to run several gravitational wave experiments simultaneously: tiny perturbations, such as oscillations of the ground generated by vehicles, can generate relative variations of the distances Sp–S_1 and Sp–S_2, which generate interference phenomena without gravitational waves. However, since the size of gravitational waves is larger than the Earth, gravitational waves would generate the sought-after interference phenomena simultaneously at each of the detectors distributed over the Earth. Only if this is observed can we claim that gravitational waves have been detected.

Exercises

4.1. Derive the relation between ω, k, and m to be satisfied by a wave solution of the form (4.3) in order to satisfy the Klein–Gordon equation (4.11).

4.2. For which value of λ does the spherically symmetric expression $\Phi(r) = \frac{\Phi_0}{r}e^{-\lambda r}$ solve the massive Klein–Gordon equation (4.11)? (This solution is a generalization of the Coulomb solution for fields corresponding to massive particles.)

Chapter 5
Electrodynamics

Classical electrodynamics is based on the generation of electromagnetic fields by charged bodies, and on the forces acting on charged bodies propagating in electromagnetic fields. These concepts are used in order to describe electron–electron scattering, as well as the scattering of two beams of electrons, at the classical level. In quantum electrodynamics, photons can be emitted and absorbed by charged particles. Within this framework, electron–electron scattering is described by the exchange of photons between the electrons. The results for the scattering angles are similar to, but slightly different from, those in classical electrodynamics. The internal angular momentum (spin) is introduced, and the Bohr atomic model is discussed.

5.1 Classical Electrodynamics

Most modern technologies—radio, television, computers, cell phones—rely on electromagnetic processes. Even though these processes are very different and mostly very complex, the fundamental laws of electrodynamics are relatively simple. The complexity of technological applications originates finally from the fact that electric or electronic components consist of an enormous number of atoms, allowing the construction of very complicated structures, circuits, chips, etc.

The fundamental laws of electrodynamics are expressed in terms of electric and magnetic fields. The electric field $\vec{E}(\vec{r}, t)$ is a vector field pointing in a certain direction. The same is true for the magnetic field $\vec{B}(\vec{r}, t)$—for instance, on the surface of the Earth, a (weak) magnetic field points towards the north pole.

The fundamental laws of electrodynamics describe how fields are induced by pointlike charges; this suffices to compute the fields induced by arbitrary distributions of charges and currents. In addition these laws imply relations between these fields, and how they act (in the form of forces) on charged objects.

The force $\vec{F}$ acting on a point-like object of velocity $\vec{v}$ and electric charge q, in a position $\vec{r}$ and at a time t, is known as the *Lorentz force*:

U. Ellwanger, *From the Universe to the Elementary Particles*,
Undergraduate Lecture Notes in Physics, DOI: 10.1007/978-3-642-24375-2_5,
© Springer-Verlag Berlin Heidelberg 2012

$$\vec{F}(\vec{r}, t) = q \left(\vec{E}(\vec{r}, t) + \vec{v} \times \vec{B}(\vec{r}, t) \right). \tag{5.1}$$

Here $\vec{v} \times \vec{B}$ denotes the so-called vector product; accordingly the force induced by a magnetic field $\vec{B}$ points into a direction perpendicular to both $\vec{B}$ and the velocity $\vec{v}$.

The equations determining the fields $\vec{E}(\vec{r}, t)$ and $\vec{B}(\vec{r}, t)$ are called the Maxwell equations. They imply that both fields satisfy the Klein–Gordon equation:

$$\left(\frac{\partial^2}{\partial t^2} - c^2 \left(\frac{\partial^2}{\partial x^2} + \frac{\partial^2}{\partial y^2} + \frac{\partial^2}{\partial z^2} \right) \right) \vec{E}(\vec{r}, t) = 0,$$

$$\left(\frac{\partial^2}{\partial t^2} - c^2 \left(\frac{\partial^2}{\partial x^2} + \frac{\partial^2}{\partial y^2} + \frac{\partial^2}{\partial z^2} \right) \right) \vec{B}(\vec{r}, t) = 0. \tag{5.2}$$

These are six equations for the six components $E_x(\vec{r}, t)$, $E_y(\vec{r}, t)$, $E_z(\vec{r}, t)$, $B_x(\vec{r}, t)$, $B_y(\vec{r}, t)$, and $B_z(\vec{r}, t)$. The wave solutions of these equations are electromagnetic waves. However, there exist no pure "electric" waves ($\vec{E}(\vec{r}, t) \neq 0$, $\vec{B}(\vec{r}, t) = 0$) or "magnetic" waves ($\vec{B}(\vec{r}, t) \neq 0$, $\vec{E}(\vec{r}, t) = 0$): an electromagnetic wave contains always both electric *and* magnetic components, since the six components above are not independent.

Instead of performing calculations involving six components that are coupled via the Maxwell equations (not given here), we can simplify life and perform all calculations in terms of four independent fields from which the fields $\vec{E}(\vec{r}, t)$ and $\vec{B}(\vec{r}, t)$ can be deduced. These independent fields are denoted by $\phi(\vec{r}, t)$ and $\vec{A}(\vec{r}, t)$, where $\vec{A}(\vec{r}, t)$ is again a three-component vector field.

The fields $\phi(\vec{r}, t)$ and $\vec{A}(\vec{r}, t)$ satisfy the Klein–Gordon equation as well,

$$\left(\frac{\partial^2}{\partial t^2} - c^2 \left(\frac{\partial^2}{\partial x^2} + \frac{\partial^2}{\partial y^2} + \frac{\partial^2}{\partial z^2} \right) \right) \phi(\vec{r}, t) = 0 \quad \text{etc.} \tag{5.3}$$

The following equations determine $\vec{E}(\vec{r}, t)$ and $\vec{B}(\vec{r}, t)$ in terms of $\phi(\vec{r}, t)$ and $\vec{A}(\vec{r}, t)$:

$$\vec{E} = -\vec{\nabla}\phi - \frac{\partial \vec{A}}{\partial t}, \tag{5.4}$$

$$\vec{B} = \vec{\nabla} \times \vec{A}. \tag{5.5}$$

We emphasize that the introduction of the fields $\phi(\vec{r}, t)$ and $\vec{A}(\vec{r}, t)$ corresponds to a simplification (since all forces and interactions are described by just four independent instead of six coupled fields), and the relations between the electric and magnetic fields are expressed implicitly by (5.4) and (5.5).

We recall that (5.3) possesses a static solution of the form

$$\phi(|\vec{r}|) = \frac{\phi_0}{r} = \frac{\phi_0}{\sqrt{x^2 + y^2 + z^2}} \tag{5.6}$$

(see (4.14) with $C=0$).

This solution describes the field $\phi(\vec{r})$ induced by a point charge q at rest at the origin. The constant ϕ_0 is related to the charge q by

$$\phi_0 = \frac{q}{4\pi\varepsilon_0}, \tag{5.7}$$

where ε_0 is the vacuum permittivity or electric field constant:

$$\varepsilon_0 \simeq 8.854 \times 10^{-12} \frac{C}{V\,m} = 8.854 \times 10^{-12} \frac{C^2\,s^2}{kg\,m^3}. \tag{5.8}$$

Since a charge at rest does not induce a field $\vec{A}(\vec{r}, t)$, (5.5) implies immediately $\vec{B}(\vec{r}, t) = 0$, and (5.4) allows the induced electric field $\vec{E}(\vec{r})$ to be computed:

$$\vec{E}(\vec{r}) = -\vec{\nabla}\phi(\vec{r}) = \frac{q}{4\pi\varepsilon_0}\frac{\vec{r}}{r^3}. \tag{5.9}$$

As a consequence of (5.1) this electric field induces a force $\vec{F}_{el}(\vec{r})$ acting on an object with electric charge q' situated at a point $\vec{r}$:

$$\vec{F}_{El}(\vec{r}) = q'\vec{E}(\vec{r}) = -q'\vec{\nabla}\phi(\vec{r}) = \frac{qq'}{4\pi\varepsilon_0}\frac{\vec{r}}{r^3}. \tag{5.10}$$

This formula can be compared to the expression (3.40) for the gravitational force acting an object with mass m in the environment of an object with mass M:

$$\vec{F}_{grav} = -\frac{mc^2}{2}\vec{\nabla}g_{00}(r) = -GmM\frac{\vec{r}}{r^3}. \tag{5.11}$$

We see that the formula for the gravitational force $\vec{F}_{grav}$ corresponds to the formula for the electric force $\vec{F}_{el}$ apart from the replacements $q \to m$, $q' \to M$, and $1/4\pi\varepsilon_0 \to G$. (These forces point in opposite directions since two electric charges of the same sign repel each other whereas the gravitational force is always attractive. For charges q and q' of opposite sign, the directions of the electric and gravitational forces coincide.) The underlying reason for this analogy is the fact that both fields ϕ and g_{00} satisfy the Klein–Gordon equation.

We should note that the four fields $\phi(\vec{r}, t)$ and $\vec{A}(\vec{r}, t)$ can be interpreted as the components of a four-vector in the four-dimensional space-time of special relativity, similar to the energy-momentum four-vector in (3.25) (see also Sect. 9.3). This allows one, in principle, to determine the electric and magnetic fields induced by a uniformly moving charge: it suffices to perform a Lorentz transformation (3.7) into a coordinate system in which the point charge has a given velocity, to transform the four fields $\phi(\vec{r}, t)$ and $\vec{A}(\vec{r}, t)$ with the *same* Lorentz transformation, and to compute subsequently $\vec{E}$ and $\vec{B}$ from (5.4) and (5.5).

This way we can determine the fields induced by an arbitrary distribution of charges and currents (i.e., moving charges), and the electromagnetic forces on carriers of electric charges from (5.1). Even though these computations can be complicated, they are based on very simple principles. Thus all electromagnetic phenomena can be traced back to these principles—except that, in the case of processes at the atomic or subatomic level, we have to take quantum effects into account.

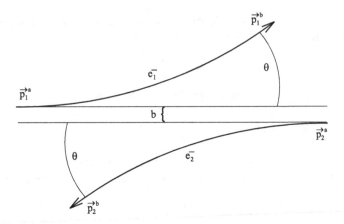

Fig. 5.1 Trajectories of two electrons with initial velocities oriented in opposite directions

5.2 Electron–Electron Scattering

Before introducing concepts of quantum electrodynamics in the next section, we will first treat at the classical level a typical process in particle physics, namely electron–electron scattering $e^- + e^- \to e^- + e^-$ as sketched in Fig. 5.1.

We denote the momenta of the electrons e_1^- and e_2^- before scattering by $\vec{p}_1^{\,a}$ and $\vec{p}_2^{\,a}$, and we assume that these vectors are oriented in opposite directions: $\vec{p}_1^{\,a} = -\vec{p}_2^{\,a}$. (This can always be achieved by the choice of an appropriate coordinate system.) The distance between the parallel lines along $\vec{p}_1^{\,a}$ and $\vec{p}_2^{\,a}$ is denoted as the *impact parameter b.*

According to (3.22), the (relativistic) energies before scattering are given by

$$E_1^a = \sqrt{m_e^2 c^4 + (\vec{p}_1^{\,a})^2 c^2} = \sqrt{m_e^2 c^4 + (\vec{p}_2^{\,a})^2 c^2} = E_2^a,$$

where m_e denotes the mass of an electron.

The momenta of the electrons after scattering will be denoted by $\vec{p}_1^{\,b}$ and $\vec{p}_2^{\,b}$, and their energies are $E_1^b = \sqrt{m_e^2 c^4 + (\vec{p}_1^{\,b})^2 c^2}$ and $E_2^b = \sqrt{m_e^2 c^4 + (\vec{p}_2^{\,b})^2 c^2}$.

The reason for characterizing the electrons by their momenta (instead of their velocities) is the conservation of total momentum (and total energy) in a scattering process: we have $\vec{p}_1^{\,a} + \vec{p}_2^{\,a} = \vec{P}_{\text{tot}} = \vec{p}_1^{\,b} + \vec{p}_2^{\,b}$ and $E_1^a + E_2^a = E_{\text{tot}} = E_1^b + E_2^b$.

Under the present assumption $\vec{p}_1^{\,a} = -\vec{p}_2^{\,a}$, this allows us to deduce $E_i^b = E_i^a$ and $|\vec{p}_i^{\,b}| = |\vec{p}_i^{\,a}|$ (for $i = 1, 2$ see Exercise 5.1).

Hence the momentum vector $\vec{p}_1^{\,a}$ of electron e_1^- before scattering differs from the momentum vector $\vec{p}_1^{\,b}$ after scattering only in its direction; the angle between these vectors is denoted as the scattering angle θ, and we have $\vec{p}_1^{\,a} \cdot \vec{p}_1^{\,b} = |\vec{p}_1^{\,a}| |\vec{p}_1^{\,b}| \cos\theta$. Due to the conservation of total momentum, the scattering angle of electron e_2^- is the same.

How is this scattering angle generated in classical electrodynamics? Electron e_1^- induces an electric field. The electron e_2^- propagates in this field in which a force $\vec{F}_2$ acts on it; this force implies a change of the direction of flight of e_2^-. The same holds after exchanging e_1^- and e_2^-. The fields and the forces are calculable, and allow a formula to be deduced for the scattering angle θ as a function of the impact parameter b (the equations below hold for the non-relativistic case $|\vec{p}_1|^2 \ll m_e^2 c^2$):

$$\tan \frac{\theta}{2} = \left(\frac{q_e^2}{4\pi\varepsilon_0} \right) \frac{m_e}{2b|\vec{p}_1|^2}, \tag{5.12}$$

where $q_e = -e$ is the charge of an electron. From this formula we find $\theta \to 0$ for $b \to \infty$, i.e., no deflection for very large distances between the trajectories. For $b \to 0$, i.e., a head-on collision, we find $\theta \to \pi$, i.e., the electrons are reflected in the directions they came from.

However, in real experiments one does not use two single electrons, but two beams of electrons. The beams collide under a very small angle (nearly head-on). Then the various electrons of one beam see the electrons of the other beam at all possible impact parameters (up to twice the diameter of the beams). Hence we find a large number of scattered electrons at all possible scattering angles.

Next, we proceed as follows: we average over all possible impact parameters b (assuming that the beams are homogeneous), and deduce a distribution $P(\theta)$ of scattered electrons. Correspondingly, $P(\theta)$ is the probability of a single electron being scattered through an angle θ. $P(\theta)$ is also known as the differential cross section. (In the literature, the differential cross section is written as $d\sigma/d\Omega$ where $d\sigma$ is the number of electrons scattered into the solid angle $d\Omega = d\varphi d(\cos\theta)$. Here we integrate over φ around the beam axis from 0 to 2π.)

The theoretical result is the *Rutherford cross section*

$$P(\theta) = \left(\frac{q_e^2}{4\pi\varepsilon_0} \right)^2 \frac{m_e^2}{16|\vec{p}_1|^4 \sin^4(\theta/2)} + (\theta \to \theta + 180°), \tag{5.13}$$

where the term "$+(\theta \to \theta + 180°)$" originates from processes in which the electrons are exchanged after scattering (see Fig. 5.2); we cannot tell which beam the detected electron came from originally. (E. Rutherford derived a similar formula for the scattering of non-relativistic electrons off heavy nuclei. Then the term "$+(\theta \to \theta + 180°)$" is absent, but the result for $P(\theta)$ is four times larger.)

This formula can be verified with the help of electron detectors placed around the interaction region. These detectors measure the number $N(\theta)$ of electrons scattered through an angle θ. Up to a multiplicative constant (depending on the density of electrons in the beams and the duration of the experiment), $N(\theta)$ should coincide with $P(\theta)$ within the statistical uncertainties or error bars.

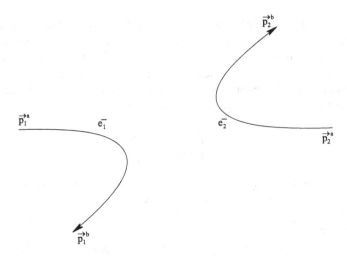

Fig. 5.2 Trajectories of two electrons with opposite initial velocities and small impact parameter b

5.3 Quantum Electrodynamics

In quantum electrodynamics the description of interactions between two charged particles differs fundamentally from classical electrodynamics: due to the duality between particles and fields we can replace the electromagnetic fields (denoted here by $\phi(\vec{r}, t)$ and $\vec{A}(\vec{r}, t)$) by a particle called the *photon*. Now the generation of an electromagnetic field by a charged particle corresponds to the emission of a photon by the particle as in Fig. 5.3, and the Lorentz force acting on a charged particle in an electromagnetic field is replaced by the absorption of a photon by the particle as in Fig. 5.4.

Now the scattering process of two electrons is described in terms of the emission of a photon by electron e_1^-, the photon being absorbed by electron e_2^-, or by the emission of a photon by electron e_2^- absorbed by electron e_1^- as in Fig. 5.5.

We have to sum over both processes in Fig. 5.5, and this sum is represented by a single *Feynman diagram* in Fig. 5.6.

Important is the fact that, in the case of an emission of a photon as in Fig. 5.3, both the total momentum and the total energy are conserved:

$$\vec{p}^a = \vec{p}^b + \vec{p}^{ph}, \quad E^a = E^b + E^{ph}. \tag{5.14}$$

The conservation of total momentum and total energy holds also for the absorption of a photon as in Fig. 5.4:

$$\vec{p}^b = \vec{p}^a + \vec{p}^{ph}, \quad E^b = E^a + E^{ph}. \tag{5.15}$$

It follows that the total momentum and the total energy remain conserved in both processes represented in Fig. 5.5:

Fig. 5.3 Emission of a
photon by a charged particle

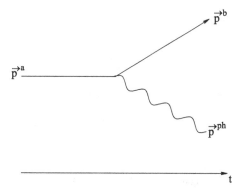

Fig. 5.4 Absorption of a
photon by a charged particle

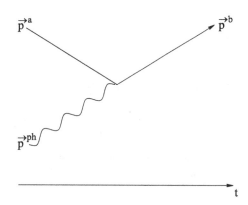

$$\vec{p}_1^a + \vec{p}_2^a = \vec{p}_1^b + \vec{p}_2^b, \quad E_1^a + E_2^a = E_1^b + E_2^b. \tag{5.16}$$

In the case $\vec{p}_1^a = -\vec{p}_2^a$ it follows as in classical electrodynamics that

$$E_i^b = E_i^a \quad \text{and} \quad \left|\vec{p}_i^b\right| = \left|\vec{p}_i^a\right| \tag{5.17}$$

for $i = 1, 2$.

Hence we find for the energy of the photon exchanged in Fig. 5.5

$$E^{ph} = 0. \tag{5.18}$$

For the modulus of the momentum of the photon we obtain

$$\left|\vec{p}^{ph}\right| = \left|\vec{p}_1^a - \vec{p}_1^b\right| = \left|\vec{p}_2^a - \vec{p}_2^b\right|. \tag{5.19}$$

Thus it differs from zero as soon as the scattering angle θ does not vanish:

$$\left|\vec{p}_1^a - \vec{p}_1^b\right| = \sqrt{\left|\vec{p}_1^a\right|^2 - 2\left|\vec{p}_1^a\right|\left|\vec{p}_1^b\right|\cos\theta + \left|\vec{p}_1^b\right|^2} \tag{5.20}$$

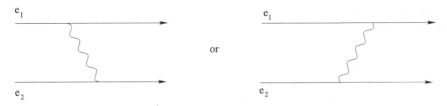

Fig. 5.5 Two processes contributing to electron–electron scattering

Fig. 5.6 Feynman diagram
combining the two processes
in Fig. 5.5

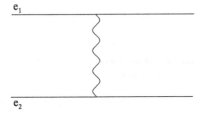

Using $\left| \vec{p}_1^{\,b} \right| = \left| \vec{p}_1^{\,a} \right|$ it follows for the momentum vector of the photon

$$\left| \vec{p}^{\,\mathrm{ph}} \right| = \left| \vec{p}_1^{\,a} \right| \sqrt{2(1 - \cos\theta)}. \tag{5.21}$$

This leads us to a paradox: according to (3.28), the energy of a photon (denoted subsequently as "classical" energy $E_{\mathrm{cl}}^{\mathrm{ph}}$) should be given, in special relativity, by

$$E_{\mathrm{cl}}^{\mathrm{ph}}(\vec{p}^{\,\mathrm{ph}}) = c \left| \vec{p}^{\,\mathrm{ph}} \right| = c \left| \vec{p}_1^{\,a} \right| \sqrt{2(1 - \cos\theta)}, \tag{5.22}$$

which differs (for $\theta \neq 0$) from the value $E^{\mathrm{ph}} = 0$ found above!

The solution of this paradox is a phenomenon specific to quantum mechanics: in quantum mechanics, the energy of a particle (such as a photon) can differ from its "classical" value given in (3.28) and (5.22) in terms of its momentum:

$$E^{\mathrm{ph}} = E_{\mathrm{cl}}^{\mathrm{ph}}(\vec{p}^{\,\mathrm{ph}}) + \Delta E^{\mathrm{ph}} \tag{5.23}$$

A particle for which the energy differs from its classical value is known as a *virtual* particle.

However, the lifetime of a virtual particle is finite: although the lifetime of an individual particle cannot be definitely predicted in quantum mechanics, we can give the probability $P(t)$ of a virtual particle (with $\Delta E \neq 0$) surviving a time t:

$$P(t) = \frac{|\Delta E|}{\hbar} \exp\left(-\frac{1}{\hbar} |\Delta E| t \right), \tag{5.24}$$

where $\hbar$ is Planck's constant already introduced in (4.9):

$$\hbar \simeq 1.055 \times 10^{-34} \, \mathrm{kg\,m^2\,s^{-1}}. \tag{5.25}$$

In view of the smallness of $\hbar$ and the rapid decrease of the exponential function, $P(t)$ is extremely small for ΔE on the order of kg m^2 s^{-2} and times t on the order of seconds. On the other hand, for energies and reaction times relevant for atomic and particle physics, the non-vanishing value of $P(t)$ allows processes such as the scattering $e^- + e^- \to e^- + e^-$ by the exchange of a virtual photon of an extremely short lifetime.

Here, and in the case of more complicated Feynman diagrams (see below), virtual particles correspond always to inner lines (and vice versa); inner lines end at interaction points, where the virtual particle is created, destroyed, or absorbs or emits another particle. These interaction points are denoted as *vertices*. External lines of Feynman diagrams correspond to particles before or after scattering for which the classical relation (3.22) between energy and momentum always holds.

Another particularity of quantum mechanics is the fact that the result of a process cannot be predicted definitely: we can only define a probability with which a given result will be obtained. In our example corresponding to the scattering $e^- + e^- \to e^- + e^-$, the "result" of the process is the scattering angle θ or, for two electron beams, the probability $P(\theta)$ of finding an electron scattered through an angle θ.

How do we compute $P(\theta)$ in quantum field theory and hence in quantum electrodynamics? Feynman diagrams not only serve to describe qualitatively a scattering process induced, in general, by the exchange of one or more virtual particles (such as the photon); a Feynman diagram also contains all the information required for the calculation of $P(\theta)$. How to translate a diagram into an algorithm for $P(\theta)$ is specified by the *Feynman rules*. In our case these rules are as follows.

(a) First we have to count the vertices at which a photon is emitted or absorbed. In the diagram in Fig. 5.6 we find two such vertices. Each vertex is associated with a coupling g known as the *coupling constant*. This constant is proportional to the charge q_e of an electron:

$$g = \frac{q_e}{\sqrt{\varepsilon_0 c \hbar}}. \tag{5.26}$$

(b) Each virtual particle has a *propagator* $\mathcal{P}(E^{\text{ph}}, \vec{p}^{\text{ph}})$ associated with it, which depends on the deviation of the energy of the particle from its classical value $c|\vec{p}^{\text{ph}}|$ [1]:

---

[1] The origin of this expression for the propagator is a solution of the Klein–Gordon equation of the form $\Phi(\vec{r}, t) = 1/(\vec{r}^2 - c^2 t^2)$ and a manipulation denoted as Fourier transformation:

$$\mathcal{P}(E, \vec{p}) = \frac{1}{\hbar} \int \frac{d^3 \vec{r} dt}{(2\pi)^2} \Phi(\vec{r}, t) e^{i(Et - c\vec{p}\vec{r})/\hbar}$$

$$= \frac{1}{\hbar} \int \frac{d^3 \vec{r} dt}{(2\pi)^2} \frac{1}{(\vec{r}^2 - c^2 t^2)} e^{i(Et - c\vec{p}\vec{r})/\hbar} = \frac{-\hbar}{(E^2 - c^2 |\vec{p}|^2)}.$$

Using the four-vector $\vec{P}_4$ introduced in (3.24, 3.25), the photon propagator can also be written elegantly as $-\hbar / \left(c^2 \left| \vec{P}_4^{\text{ph}} \right|^2 \right)$.

$$P(E^{\text{ph}}, \vec{p}^{\text{ph}}) = \frac{-\hbar}{\left(E^{\text{ph}\,2} - c^2 \left|\vec{p}^{\text{ph}}\right|^2\right)}. \tag{5.27}$$

We have seen that, owing to energy and momentum conservation at the vertices, the energy and the momentum of the photon are determined by the energies and momenta of the electrons before and after scattering; for the diagram corresponding to Fig. 5.6 we obtained $E^{\text{ph}} = 0$ and $\left|\vec{p}^{\text{ph}}\right| = \left|\vec{p}_1^{\text{a}}\right| \sqrt{2(1 - \cos\theta)}$. These results have to be used in the photon propagator, which will appear as a factor in the final result.

(c) Up to a prefactor $1/(256\pi^2 m_{\text{e}}^2)$ (under the present hypothesis $|\vec{p}_i|^2 \ll m_{\text{e}}^2 c^2$) the sought-after expression for $P(\theta)$ is the square of a quantity denoted as the amplitude $A(\theta)$ in quantum mechanics. $A(\theta)$ is essentially the product of all previous factors: a power of g for each vertex, a propagator for each virtual particle (here: the photon), and a factor depending on the masses and momenta of the "external" particles. For $|\vec{p}_i|^2 \ll m_{\text{e}}^2 c^2$ this factor is given by $4m_{\text{e}}^2 c^3 \hbar$.

Hence we obtain for the amplitude $A(\theta)$ (with $E^{\text{ph}} = 0$ in the propagator of the photon)

$$A(\theta) = 4m_{\text{e}}^2 c^3 \hbar g^2 P(E^{\text{ph}}, \vec{p}^{\text{ph}}) = \frac{4m_{\text{e}}^2 g^2 \hbar^2 c}{\left|\vec{p}^{\text{ph}}\right|^2}. \tag{5.28}$$

Using (5.26) for g and (5.21) for $\left|\vec{p}^{\text{ph}}\right|$ we finally obtain for $P(\theta)$

$$P(\theta) = \frac{1}{256\pi^2 m_{\text{e}}^2} A^2(\theta) = \left(\frac{m_{\text{e}} q_{\text{e}}^2}{8\pi\varepsilon_0 \left|\vec{p}_1^{\text{a}}\right|^2 (1 - \cos\theta)}\right)^2, \tag{5.29}$$

which coincides with the classical result (5.13) because $1 - \cos\theta = 2\sin^2(\theta/2)$. Hence, up to now, the formalism of quantum electrodynamics gives the same result for the probability $P(\theta)$ for $e^- + e^- \to e^- + e^-$ as classical electrodynamics. The terms $(\theta \to \theta + 180°)$ remain to be added, leading, however, to a different result: these terms are to be added in the amplitude $A(\theta)$ (with a negative sign due to Fermi statistics), whereupon we obtain for $P(\theta)$

$$P(\theta) = \left(\frac{q_{\text{e}}^2}{4\pi\varepsilon_0}\right)^2 \frac{m_{\text{e}}^2}{16|\vec{p}_1|^4}$$
$$\times \left(\frac{1}{\sin^4(\theta/2)} + \frac{1}{\cos^4(\theta/2)} - \frac{1}{\sin^2(\theta/2)\cos^2(\theta/2)}\right), \tag{5.30}$$

the *Mott cross section* (given here in the non-relativistic limit). The additional terms lead to a measurable deviation of $P(\theta)$ from the classical result (5.13).

Furthermore, the process represented in Fig. 5.6 is not the only one contributing to the probability $P(\theta)$ in quantum electrodynamics. Since an electron can emit a

Fig. 5.7 Feynman diagram corresponding to the exchange of two photons

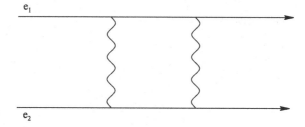

photon, it can also emit two photons one after the other. Likewise it can absorb two photons. Such processes can be represented by somewhat more complicated Feynman diagrams as in Fig. 5.7.

How do we take into account the contributions to $P(\theta)$ corresponding to such diagrams? Each diagram corresponds to an expression (an amplitude $A_i(\theta)$) calculable following the Feynman rules. The total amplitude is given by $A_{tot}(\theta) = \sum_i A_i(\theta)$, and the total probability $P_{tot}(\theta)$ by $P_{tot}(\theta) = A_{tot}^2(\theta)/(256\pi^2 m_e^2)$.

We can estimate the order of magnitude of the contributions to A_{tot} originating from more complicated diagrams with the help of a consideration relying exclusively on the number of vertices in the diagrams. First we define the so-called fine structure constant α, which depends on the coupling g (i.e., on the charge $q_e = -e$ of the electron):

$$\alpha = \frac{g^2 \hbar}{4\pi} = \frac{e^2}{4\pi\varepsilon_0 \hbar c}. \tag{5.31}$$

The advantage of this quantity is that it is dimensionless, i.e., a pure number calculable using the known values of e, ε_0, $\hbar$ and c:

$$\alpha \simeq \frac{1}{137}. \tag{5.32}$$

Since the number of powers of g in an amplitude $A_i(\theta)$ is given by the number N_V of vertices, the contributions to an amplitude are always proportional to $\alpha^{N_V/2}$. Assuming that the other factors contributing to $A_i(\theta)$ are of the same order (which holds due to the denominator 4π in (5.31)), it follows from the smallness of α that the contribution of a given diagram decreases with the number N_V of vertices. Accordingly the numerically most important contribution to the total amplitude originates from the diagram with the smallest number of vertices (with at least one exchanged photon)—this is the contribution computed above. The contributions from diagrams such as in Fig. 5.7 are smaller by about a factor α.

Even if in most cases the contributions of diagrams with more vertices are relatively small, they are non-vanishing and lead to measurable corrections. These corrections have been compared to results of measurements, and the formalism of quantum electrodynamics has been confirmed to high precision.

In general, the treatment of photons and electrons is very similar in quantum field theory: both are described by fields, and the square of the fields (depending

Fig. 5.8 Annihilation of a
particle with its antiparticle
into a photon

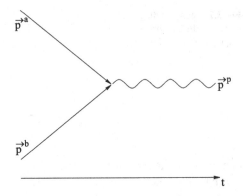

on the position $\vec{r}$ and the time t) can be interpreted as the probability of finding a
corresponding particle at the position $\vec{r}$ at the time t.

The different treatment of electrons and photons above is due to the considered
process, the scattering of electrons by the exchange of a virtual photon.

On the other hand, the scattering of photons by the exchange of electrons (or
positrons) is possible as well. In quantum field theory each process for which at
least one corresponding Feynman diagram can be found is, in principle, possible.
We can indeed draw such diagrams if we use vertices that can be obtained from
those of Figs. 5.3 and 5.4 by a simple manipulation: in Fig. 5.3 we can reverse the
chronological direction of the line associated with $\vec{p}^{\,b}$, which gives the vertex in
Fig. 5.8. In quantum field theory, the time reversal of a line implies the replacement
of the corresponding particle by its antiparticle. Accordingly the vertex in Fig. 5.8
describes the annihilation of an electron and a positron into a photon.

Likewise, we can reverse the chronological direction of the line of the electron with
momentum $\vec{p}^{\,a}$ in Fig. 5.4, leading to the vertex in Fig. 5.9. This vertex describes the
decay of a photon into an electron and a positron. (Owing to energy and momentum
conservation, at least one of the lines in Figs. 5.8 and 5.9 must correspond to a virtual

Fig. 5.9 Decay of a photon
into a particle and an
antiparticle

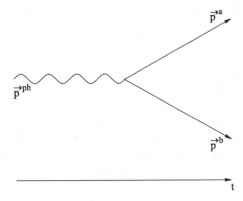

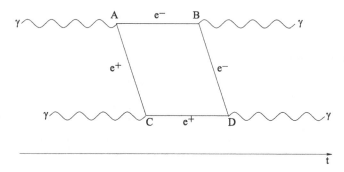

Fig. 5.10 Diagram contributing to photon–photon scattering

particle, i.e., an inner line of a Feynman diagram for which the classical relation (3.22) between energy and momentum does not hold.)

With the help of the vertices in Figs. 5.8 and 5.9 we can draw, e.g., the diagram in Fig. 5.10 (or other chronological orders of the vertices A, B, C, and D). This diagram describes photon–photon scattering $\gamma + \gamma \rightarrow \gamma + \gamma$ (or light-by-light scattering) by the exchange of electrons or positrons.

This phenomenon is something new that would not be possible in classical electrodynamics: if two light rays cross each other, in classical electrodynamics the two rays penetrate each other without a single photon being scattered. However, because of the diagram in Fig. 5.10, photon–photon scattering can take place in quantum electrodynamics, whereupon some of the photons are scattered through various scattering angles θ (like electrons in the case of electron–electron scattering discussed above). Even though the probability of such a process is very small, since the probability is proportional to α^4 (with α given in (5.32)), the phenomenon of photon–photon scattering has been verified with the help of very intense laser beams.

The fact that the number of electrons or positrons connected to the vertices of Figs. 5.3, 5.4, 5.8, and 5.9 is even (in contrast to the number of photons) has its origin (among other things, see the next section) in the conservation of the electric charge: the sum of the electric charges of the particles before a process (emission or absorption) is always equal to the sum of the electric charges of the particles after the process, similar to the conservation of energy and momentum.

5.4 Internal Angular Momentum

Before discussing the internal angular momentum, we want to recall the definition of the external angular momentum: a particle with mass m, rotating around an axis on a circular orbit with velocity $\vec{v}$ as in Fig. 5.11, carries an angular momentum $\vec{L} = m\,\vec{r} \times \vec{v}$ oriented along the axis. (The direction of $\vec{L}$ along the rotation axis follows from the directions of the vectors $\vec{r}$ and $\vec{v}$ according to the corkscrew rule.)

Fig. 5.11 Definition of the external angular momentum $\vec{L}$ of a particle in terms of its position $\vec{r}$ and its velocity $\vec{v}$

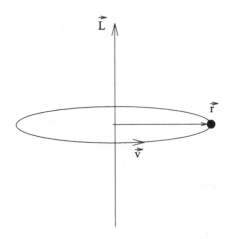

If a body is rotating around an axis passing through its center, we can associate an internal angular momentum with it. For instance, a sphere of homogeneous density, total mass M, and radius R possesses an internal angular momentum

$$L = \frac{2}{5}MRv, \tag{5.33}$$

where v is the rotational velocity of its equator. A planet rotating around the Sun possesses an orbital angular momentum as well as an internal angular momentum arising from its rotation around its own axis. Angular momenta are conserved in the absence of external forces, or in cases of so-called central forces (such as the gravitational force generated by the Sun), which explains the stability of the orbits of planets.

In fact, elementary particles carry internal angular momenta as well. This can be proven, e.g., by their decays, where the total angular momentum remains conserved. The internal angular momentum is denoted as *spin*.

At first it was tempting to apply the idea of a rotating massive sphere to elementary particles as well. However, for various reasons this idea cannot be correct:

- massless particles (such as photons and, approximately, neutrinos) also carry a spin;
- we cannot associate a radius R with elementary particles;
- the modulus of the spin of a given species of particles is always an integer or half-integer multiple of Planck's constant $\hbar$ known from (4.9) and (5.25); $\hbar$ possesses the same units as angular momentum. The modulus of the spin of a given species of particle is always the same and cannot be changed, e.g., by the emission or absorption of a photon.

For these reasons a formula of the kind (5.33) cannot be applied to the spin. How should we imagine the spin, then? To this end it is necessary to use the representation of particles in the form of waves of the corresponding field.

Fig. 5.12 Sketch of a vector field rotating around the direction of propagation of a wave

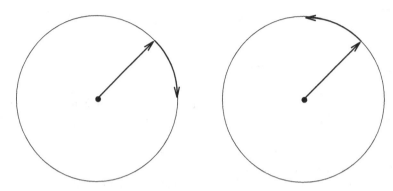

Fig. 5.13 Sketch of a vector field rotating clockwise or anticlockwise around the direction of propagation of a wave, viewed along the direction of propagation

Fields can be vector fields pointing in a given direction. In the case of a propagating wave of a vector field, this direction can vary along the direction of propagation, as sketched in Fig. 5.12. The small arrows in Fig. 5.12 denote the direction of the field, which is always oriented perpendicular to the direction of propagation for massless fields: these so-called polarization vectors rotate around the axis corresponding to the direction of propagation.

If we look along the beam axis, the polarization vector can rotate clockwise or anticlockwise, as in Fig. 5.13. In the case of light waves (waves of electromagnetic fields, which are vector fields) we refer to right-handed or left-handed polarization. (Linearly polarized light corresponds to a superposition of right-handed and left-handed polarized waves.)

The corresponding particles are photons, with which we can indeed associate a spin $\hbar$. Generally a particle with a spin that is an integer multiple of $\hbar$ (or vanishes) is denoted as a *boson*.

Electrons and positrons, on the other hand, carry spin $\hbar/2$. Besides electrons and positrons there exist more particle species with spin $\hbar/2$: the quarks, and more leptons (see Chap. 7), such as the muon. Particles with spin $\hbar/2$ (or, generally, odd-integer multiples of $\hbar/2$) are denoted as *fermions*.

Fermions possess the following property: their number is either conserved, or they are created or annihilated as a particle–antiparticle pair in every process. In contrast to fermions, bosons can be created or annihilated without additional antiparticles. If they are neutral (as the photon), they can be their own antiparticles.

Generally, matter (quarks and leptons) consists of fermions, whereas carriers of interactions such as the photon (and more, see the following chapters) are bosons.

Strictly speaking, the Feynman rules discussed above—the expressions for the electron–photon vertices, the photon propagator, and the treatment of the incoming and outgoing electron and positron lines—depend on the directions of the spin of the participating particles. However, the rules given above are valid for the cases considered here, where we average over the directions of the spins of all participating particles.

5.5 The Bohr Atomic Model

In the early days of the development of quantum mechanics, Bohr considered a simple model for atoms where electrons revolve around the nucleus like planets around the Sun. Here we will confine ourselves to the simplest atom, the hydrogen atom, where a single electron rotates around a nucleus consisting of a single proton, and assume that the orbit is a circle of radius r.

The attractive force exerted by the proton on the electron is given by (5.10) with $q = e$, $q' = -e$, and hence of modulus

$$\left| \vec{F}_{El} \right| = \frac{e^2}{4\pi\varepsilon_0 r^2}. \tag{5.34}$$

This force has to compensate the centrifugal force of the electron given by

$$\left| \vec{F}_{centrif} \right| = \frac{m_e v^2}{r}, \tag{5.35}$$

where m_e is the mass and v the velocity of the electron. Requiring (5.34) and (5.35) to coincide, we can derive a formula for v depending on the radius r:

$$v^2 = \frac{e^2}{4\pi\varepsilon_0 m_e r} \tag{5.36}$$

Now we consider the modulus of the orbital momentum of the electron $L = m_e v r$, the internal angular momentum (spin) of the electron does not play a role here. Bohr made the hypothesis that L can assume only integer multiples of $\hbar$:

$$L = m_e v r = n\hbar, \tag{5.37}$$

where n is an integer larger than 0. From this hypothesis and (5.36) it follows that r can assume only particular values:

$$r(n) = n^2 R_B, \tag{5.38}$$

where R_B is known as the Bohr radius:

$$R_B = \frac{4\pi\varepsilon_0 \hbar^2}{e^2 m_e} = 0.52917 \times 10^{-10}\,\mathrm{m} = 0.52917\,\text{Å}. \tag{5.39}$$

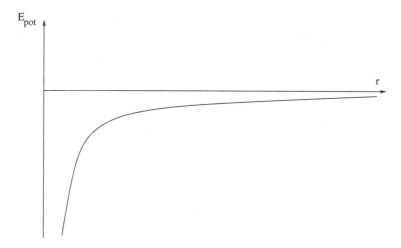

Fig. 5.14 Schematic representation of the r dependence of the electromagnetic potential energy generated by a nucleus at the origin

This explains the typical diameter of atoms of a few angstroms.

What are the consequences for the total energy of the electron? The total energy of the electron is the sum of its kinetic energy $E_{\text{kin}} = \frac{1}{2}mv^2$ and its potential energy $E_{\text{pot}}(r)$. The dependence of the potential energy on the radius r of the orbit is given by the fact that the derivative of $E_{\text{pot}}(r)$ with respect to r gives the force (the sign of $E_{\text{pot}}(r)$ follows from the direction of the force):

$$E_{\text{pot}}(r) = -\frac{e^2}{4\pi\varepsilon_0 r}. \tag{5.40}$$

$E_{\text{pot}}(r)$ is represented schematically in Fig. 5.14.

If we replace the allowed values depending on n for v in E_{kin} as well as for r in E_{pot}, we obtain for the total energy $E_{\text{tot}} = E_{\text{kin}} + E_{\text{pot}}$

$$E_{\text{tot}}(n) = -\frac{1}{n^2}E_{\text{R}}, \quad E_{\text{R}} = \frac{m_e e^4}{32\pi^2\varepsilon_0^2\hbar^2}. \tag{5.41}$$

E_{R} is known as the Rydberg energy. The possible values for the total energy $E_{\text{tot}}(n)$ of an electron can be determined very well, since an electron can jump only from one allowed orbit (corresponding to a given value of n) to another allowed orbit (of lower total energy, i.e., a lower value of n), thereby emitting photons, whose energies can be measured very well. These measurements—the results of which were known before the Bohr atomic model—agree very well with the allowed orbits, i.e., energies, of electrons according to this model.

Bohr motivated the assumption (5.37) by the relation (4.10) between the momentum p of a particle and the corresponding wavelength λ, $p = 2\pi\hbar/\lambda$: if

Fig. 5.15 Sketch of a wave
corresponding to an electron
turning around a nucleus

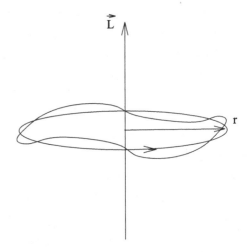

one describes the rotating electron by a wave along the orbit as in Fig. 5.15, the wave
should "chase its own tail": after each turn a wave crest should encounter a wave
crest, and a wave trough a wave trough.

This implies that the circumference $2\pi r$ of a circular orbit must be an integer
multiple n of the wavelength λ:

$$2\pi r = n\lambda = n\frac{2\pi\hbar}{p}. \tag{5.42}$$

Using $L = rp = rm_e v$, Bohr's hypothesis (5.37) follows. In 1922 Niels Bohr was
honored with the Nobel prize for his contributions to the development of quantum
mechanics.

Exercises

5.1. Consider the electron–electron scattering process in Fig. 5.1 assuming $\vec{p}_1{}^a = -\vec{p}_2{}^a$. Derive $E_i^b = E_i^a$ and $|\vec{p}_i{}^b| = |\vec{p}_i{}^a|$ for $i = 1, 2$ from the conservation of
momentum and energy.
5.2. Compute the energy and the wavelength of a photon emitted by an electron
of a hydrogen atom that jumps from an orbit with $E_{\text{tot}}(n = 2)$ into an orbit with
$E_{\text{tot}}(n = 1)$ (use (5.41)).

Chapter 6
The Strong Interaction

The strong interaction is responsible for the binding of protons and neutrons in nuclei, and for the binding of quarks inside protons and neutrons. It is generated by the exchange of gluons between quarks. Quarks carry a new quantum number known as color. The strong force acting on quarks inside protons and neutrons prohibits their separation beyond distances larger than the diameters of protons and neutrons. This phenomenon explains why free quarks are not observed, but that quarks are confined in protons, neutrons, and various additional bound states denoted as baryons or mesons, which together are known as hadrons. Gluons carry color as well, and are likewise confined inside hadrons.

6.1 Quantum Chromodynamics

We have seen in the introduction that the strong interaction is responsible for the attractive force between quarks, the constituents of protons and neutrons. The attractive force between two protons or a proton and a neutron, etc. is just a secondary effect of this fundamental force between quarks.

In quantum field theory, the electromagnetic interaction (the force between two charged objects) is generated by the exchange of one or more photons. Likewise the strong interaction is generated by the exchange of particles with spin $\hbar$, the *gluons*, as illustrated in Fig. 6.1.

We recall the Feynman rule according to which the electron–photon vertex (see Figs. 5.3, 5.4, 5.8, and 5.9) is proportional to the electron charge q_e, see (5.26). In the case of the strong interaction, the quark–gluon vertex is proportional to a strong charge q^s of the quarks, which is independent of its electric charge. There exist, however, *three* strong charges q_i^s, $i = 1, 2, 3$, which are also denoted as *colors*. (For this reason this theory is also called *quantum chromodynamics* or QCD.) This color changes whenever a gluon is emitted or absorbed by a quark, as in Fig. 6.2.

Since we have $q_j^s \neq q_i^s$, the gluons carry "strong charges", i.e., colors, as well. (For this reason they are denoted by G_{ij} in Fig. 6.2.) The color of a quark can be

U. Ellwanger, *From the Universe to the Elementary Particles*,
Undergraduate Lecture Notes in Physics, DOI: 10.1007/978-3-642-24375-2_6,
© Springer-Verlag Berlin Heidelberg 2012

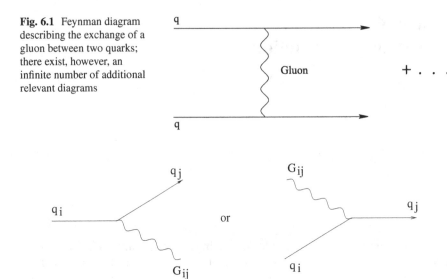

Fig. 6.1 Feynman diagram describing the exchange of a gluon between two quarks; there exist, however, an infinite number of additional relevant diagrams

Fig. 6.2 Emission or absorption of a gluon by a quark

represented by a three-component vector $\vec{q}^{\,s}$ in "color space". After the emission or absorption of a gluon by a quark the direction of this vector has changed, but not its modulus.

The result of a calculation of a probability as $P(\theta)$ (as, e.g., in (5.29)) depends only on the modulus $q^s = |\vec{q}^{\,s}|$ of the color vector of quarks, which remains invariant. (If all observables are independent of a variable such as the direction of the vector $\vec{q}^{\,s}$ in color space, we talk about a *symmetry* with respect to variations of this variable. The components of the vector $\vec{q}^{\,s}$ are complex numbers in general. In the case of three-component complex vectors with constant modulus, this symmetry is denoted by $SU(3)$, see Chap. 9.)

In analogy to (5.31) we define a strong fine structure constant α_s:

$$\alpha_s = \frac{(q^s)^2}{4\pi\varepsilon_0\hbar c}. \tag{6.1}$$

The small value $\alpha \simeq 1/137$ of the electromagnetic fine structure constant signifies that this interaction is relatively weak. In particular this small value implies that contributions to a given process from Feynman diagrams with a larger number of vertices are relatively small and normally negligible.

In the case of the strong interaction, the numerical value of the fine structure constant is

$$\alpha_s \simeq 1. \tag{6.2}$$

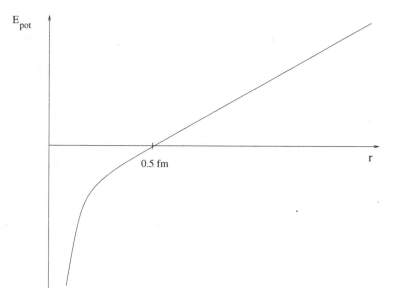

Fig. 6.3 Schematic representation of the r dependence of the potential energy induced by the strong interaction (the exchange of an infinite number of gluons) between two quarks

Consequently this interaction is relatively strong, and contributions to the interaction between two quarks from more complex Feynman diagrams (with more vertices) are not negligible. There exist an infinite number of such diagrams, and the exact behavior of the strong interaction has not been computed to date. (Nowadays high-performance computers are used for this purpose, whose architecture is adapted to this objective.)

The most important effect of diagrams with more vertices concerns the dependence of the strong force on the distance r between two quarks. The r dependence of the electric force between two charged particles is given in (5.10) and (5.34), according to which it decreases as $1/r^2$.

In the case of the strong interaction, we find, on the other hand, that the modulus of the attractive force is approximately independent of r for $r \gtrsim 0.5\,\mathrm{fm} = 0.5 \times 10^{-15}$ m. For such values of r its numerical value is

$$\left| \vec{F}_{\mathrm{strong}} \right| \sim 1.8 \times 10^5 \,\mathrm{kg\,m\,s}^{-2}. \tag{6.3}$$

Accordingly, also the expression for the potential energy $E_{\mathrm{pot}}(r)$ differs from (5.40): for $r \gtrsim 0.5$ fm it behaves as

$$E_{\mathrm{pot}}(r) = r \left| \vec{F}_{\mathrm{strong}} \right|. \tag{6.4}$$

This behavior of $E_{\mathrm{pot}}(r)$ is sketched in Fig. 6.3.

We recall that, in the absence of external forces, the total energy $E_{tot} = E_{pot} + E_{kin}$ of a system is conserved. If E_{pot} increases with r as in (6.4) and in Fig. 6.3, the maximally possible distance between two quarks corresponds to the maximally possible value of E_{tot}, given by $E_{tot} = E_{pot}(r_{max})$ and $E_{kin} = 0$. However, on average E_{pot} and E_{kin} are of the same order.

Let us consider the orders of magnitude of these energies for quarks inside a proton. Given a distance of about 0.5 fm between two quarks the mean value of the potential energy is, according to (6.3) and (6.4),

$$E_{pot} \sim 0.5 \times 10^{-15}\,\text{m} \ \times \ 1.8 \times 10^5\,\text{kg m s}^{-2} \sim 0.9 \times 10^{-10}\,\text{kg m}^2\,\text{s}^{-2}. \quad (6.5)$$

For a rough estimate we can assume that the mean velocity of a quark inside a proton is on the order of the speed of light c. The mass of a quark is about a third of the mass of a proton. Correspondingly the mean value of the kinetic energy of both quarks is on the order of $2 \times \frac{1}{2} \frac{m_p}{3} c^2 = \frac{1}{3} m_p c^2$. With $m_p \simeq 1.7 \times 10^{-24}$ g and $c \simeq 3 \times 10^8$ m s^{-1} this leads to

$$E_{kin} \sim 0.5 \times 10^{-10}\,\text{kg m}^2\,\text{s}^{-2}. \quad (6.6)$$

Thus it is indeed of the same order as the potential energy (6.5).

On the one hand this calculation explains (for given quark masses) the order of the radius of a proton, i.e., the average distance between two quarks. On the other hand we see the difficulties to be overcome if we want to separate two quarks by a much greater distance: the necessary potential energy must be much larger, which rapidly becomes impossible: in order to separate two quarks by 1 mm $= 10^{12}$ fm we would need 10^{12} times the energy available in a proton (or neutron)! (In contrast, the separation of an electron from a positron or a nucleus requires just a finite energy since, for $r \to \infty$, the electric potential energy approaches a constant, as in Fig. 5.14.)

The impossibility of separating quarks to distances much larger than a fermi is denoted as *confinement*. It follows that free quarks do not exist. Quarks are always bound, either

(a) in configurations of three quarks carrying three different colors, corresponding to a "white" or color-neutral state, or
(b) in configurations of one quark and one antiquark of opposite colors, which also results in a color-neutral or "white" state.

6.2 Bound States of Quarks

These bound states of quarks are denoted as *hadrons*, the three-quark states as *baryons* and the quark–antiquark states as *mesons*. The proton (three quarks uud) and the neutron (three quarks ddu) are members of the family of baryons. The spin of a baryon is always a half-integer multiple of $\hbar$; $\hbar/2$ in the case of protons and neutrons.

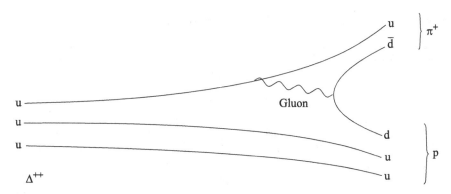

Fig. 6.4 Description of the decay of the baryon Δ^{++} into a proton p and a pion π^+ at the level of quarks and gluons

There also exist baryons consisting of three u quarks (denoted as Δ^{++}) or three d quarks (Δ^-) with spin $3\hbar/2$ and about 1.3 times as heavy as a proton.

The most important mesons are the pions π^+, π^0, and π^-, consisting of the following quarks:

$$\pi^+ : u\bar{d}$$
$$\pi^0 : u\bar{u} + d\bar{d}$$
$$\pi^- : d\bar{u} \tag{6.7}$$

Here $\bar{u}$ denotes an anti-u quark, the antiparticle of a u quark with electric charge $-\frac{2}{3}e$, and $\bar{d}$ an anti-d quark with charge $+\frac{1}{3}e$. (This explains the electric charges of the pions.) The pions carry spin 0, and due to a large binding energy (see (1.7) for the mass of a nucleus) they are relatively light: $m_{\pi^+} \sim m_{\pi^0} \sim m_{\pi^-} \sim m_p/6$ (π^- is the antiparticle of the pion π^+ of the same mass). It follows from $m_{\Delta^{++}} > m_p + m_{\pi^+}$ that the baryon Δ^{++} can decay into a proton and a pion, as illustrated in Fig. 6.4.

Accordingly the baryons Δ^{++} are unstable; owing to the strength of the strong interactions the probability of the decay of a Δ^{++} baryon is very large, and its mean lifetime τ_Δ is very small: $\tau_\Delta \sim 5.2 \times 10^{-24}$ s. (We can hardly speak of a "particle"; often the notion "resonance" is used.) We can detect such an unstable particle by its decay products: in Fig. 6.4 the sum of the momenta $\vec{p}_\pi$ of the pion and $\vec{p}_p$ of the proton is equal to the momentum $\vec{p}_\Delta$ of the baryon Δ^{++}, and the sum of the energies $E_\pi + E_p$ is equal to E_Δ. If, in an experiment, we find a large number of pions and protons whose momenta and energies satisfy the relation $(E_\pi + E_p)^2 - (\vec{p}_\pi + \vec{p}_p)^2 c^2 = m_\Delta^2 c^4$ for a given value of m_Δ, we can conclude that particles Δ^{++} of corresponding mass had been produced, even if they decayed immediately thereafter. (In fact even the pions π^+ are unstable due to the weak interaction, see the next chapter.)

Before we give more precise values for the masses of these particles, it is helpful to introduce more convenient units than grams or kilograms for particle masses (elementary or composite).

Table 6.1 Masses and electric charges (in multiples of e) of the known quarks

Quark	u	d	s	c	b	t
Masses [GeV/c^2:]	~ 0.3	~ 0.3	~ 0.5	~ 1.4	~ 4.4	~ 173
Electric charge:	$+\frac{2}{3}$	$-\frac{1}{3}$	$-\frac{1}{3}$	$+\frac{2}{3}$	$-\frac{1}{3}$	$+\frac{2}{3}$

First we use, as a unit of energy, $1\,\text{eV} = 1$ electron volt, which corresponds to the energy gained by a particle of electric charge e on passing through an electric potential difference of 1 volt. 1 volt is equal to $1\,\text{J}\,\text{C}^{-1}$, where $1\,\text{J} = 1\,\text{joule} = 1\,\text{kg}\,\text{m}^2\,\text{s}^{-2}$ and C stands for Coulomb. Using $e \simeq 1.6 \times 10^{-19}\,\text{C}$ we find

$$1\,\text{eV} \sim 1.6 \times 10^{-19}\,\text{J}. \tag{6.8}$$

Now we specify masses in multiples of eV/c^2, where c is the speed of light. (This allows us immediately to obtain the energy in electron volts "stored" in a mass m via the formula $E = mc^2$.) Expressed in kilograms, this unit is

$$1\,\text{eV}/c^2 \simeq \frac{1.6 \times 10^{-19}\,\text{J}}{9 \times 10^{16}\,\text{m}^2\,\text{s}^{-2}} \simeq 1.78 \times 10^{-36}\,\text{kg}. \tag{6.9}$$

As usual, $1\,\text{keV} = 10^3\,\text{eV}$, $1\,\text{MeV} = 10^6\,\text{eV}$, and $1\,\text{GeV} = 10^9\,\text{eV}$.

The masses of the pions and some baryons are

$$m_{\pi^\pm} \simeq 139.6\,\text{MeV}/c^2,$$
$$m_{\pi^0} \simeq 135.0\,\text{MeV}/c^2,$$
$$m_{\text{p}} \simeq 0.938\,\text{GeV}/c^2,$$
$$m_{\text{n}} \simeq 0.939\,\text{GeV}/c^2,$$
$$m_{\Delta^{++}} \sim m_{\Delta^-} \sim 1.23 \ \text{GeV}/c^2. \tag{6.10}$$

All these hadrons consist of u and d quarks with masses of about $300\,\text{MeV}/c^2$. (The mass of a particle, such as a quark, that is never observed free cannot be measured directly and is thus ill defined. Usually we determine the mass of a particle by the relation $E^2 = m^2c^4 + \vec{p}^2c^2$ and independent measurements of the energy E and the momentum $\vec{p}$, which is impossible for confined quarks. The value $\sim 300\,\text{MeV}/c^2$ is used in so-called potential models, which describe quite well the measured masses of hadrons.)

There exist additional heavier quarks, which are unstable due to the weak interaction (see the next chapter). All of them carry spin $\hbar/2$ and they are denoted as the s quark (s for "strange"), c quark (c for "charm"), b quark (b for "bottom"), and t quark (t for "top"). All quarks carry color, and their masses and electric charges (in multiples of the elementary charge e) are given in Table 6.1.

All quarks form hadrons (apart from the top quark, which decays too fast), in particular mesons (with integer spin) consisting of a quark q and an antiquark $\bar{\text{q}}$: the mesons K^+, K^-, K^0, and $\bar{K}^0$ (corresponding to $u\bar{s}$, $\bar{u}s$, $d\bar{s}$, and $\bar{d}s$, respectively);

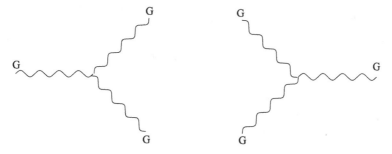

Fig. 6.5 Emission and absorption of gluons by gluons

Fig. 6.6 The four-gluon vertex

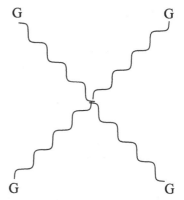

ω (s$\bar{\text{s}}$); J/Ψ (c$\bar{\text{c}}$); Υ (b$\bar{\text{b}}$); and more, with masses of about $m_{\text{meson}} \sim m_{\text{q}} + m_{\bar{\text{q}}}$. The quarks u, d, and s alone form about 100 different hadrons. The 1969 Nobel prize was awarded to Murray Gell–Mann for the relatively simple description of all these hadrons in the quark model [11].

We recall that the emission or absorption of a gluon changes the color of a quark (see Fig. 6.2) and, correspondingly, the gluons carry colors as well. Hence the gluons carry also a strong charge—unlike photons, which carry no electric charge. Owing to their strong charge, gluons can emit or absorb other gluons, as shown in Fig. 6.5. There even exists a four-gluon vertex, as illustrated in Fig. 6.6.

Accordingly there exist Feynman diagrams contributing to gluon–gluon scattering as in Fig. 6.7 and, as in the case of quark–quark scattering, an infinite number of diagrams with more vertices that are not negligible.

These diagrams generate an attractive force, implying confinement of gluons, just as for quarks. ("Confined" gluons exist inside hadrons, where their exchange between quarks generates the strong attractive force.)

There probably exist bound (but very unstable) states of masses of about $1.5\,\text{GeV}/c^2$ consisting only of gluons, called "glueballs". However, owing to their extremely short lifetime, among other reasons, these states are very difficult to detect.

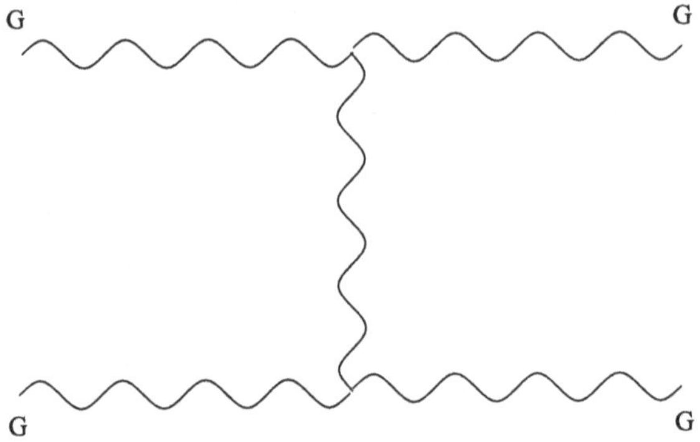

Fig. 6.7 Feynman diagram contributing to gluon–gluon scattering

6.3 Summary

At first sight the strong interaction is very similar to quantum electrodynamics: like quantum electrodynamics, it is generated by the exchange of massless particles of spin $\hbar$, the gluons.

For the following reasons the strong interaction differs from the electromagnetic force, however:

(a) Instead of a "single" electric charge there exist three strong charges, the three colors. Only quarks, not electrons or additional leptons, carry strong charges, i.e., colors.
(b) The numerical value of the strong fine structure constant α_s is about one, much larger than $\sim 1/137$. Accordingly the force is stronger, in particular its behavior at larger distances is very different from that of the electric force. This implies the confinement of quarks, according to which the only observable states are "white" (or "color neutral"), i.e., baryons, consisting of three quarks (qqq), and mesons, consisting of a quark and an antiquark ($q\bar{q}$). Between hadrons, the strong interaction leads also to attractive forces, which are responsible for the binding of protons and neutrons in nuclei, but these forces decrease rapidly with increasing distance.
(c) Gluons carry a strong charge and feel the strong force, and are also bound in hadrons. In contrast to photons they are not observed as free particles.

Exercise

6.1. The following baryons with corresponding masses consist only of u, d, and s quarks:

neutron ($0.939\,\mathrm{GeV}/c^2$), proton ($0.938\,\mathrm{GeV}/c^2$), Λ^0($1.116\,\mathrm{GeV}/c^2$), Σ^+($1.189\,\mathrm{GeV}/c^2$), Σ^0($1.193\,\mathrm{GeV}/c^2$), Σ^-($1.197\,\mathrm{GeV}/c^2$), Ξ^0($1.315\,\mathrm{GeV}/c^2$), Ξ^-($1.321\,\mathrm{GeV}/c^2$).

Estimate their quark content from their charges and mass differences.

Chapter 7
The Weak Interaction

The weak interaction transforms different quark species into each other. Today we know about six species of quarks, which differ in their masses and electric charges. Leptons are, by definition, those elementary particles that feel the weak but not the strong interaction: the electron, muon, tau-lepton, and their three corresponding nearly massless neutrinos, which can be generated by the weak interaction. The weak interaction is based on the exchange of very massive W and Z particles. This explains why the processes of the weak interaction are relatively rare. The masses of the W and Z particles require the introduction of the so-called Higgs field, which implies the existence of the Higgs boson, which is still being searched for in today's experiments at particle accelerators. Owing to its dependence on the orientation of the spin of the particles, the weak interaction violates mirror symmetry (parity): if our world were observed through a mirror, the behavior of the weak interaction would be different. Finally, even particle–antiparticle symmetry (so-called CP symmetry) is violated. Recent observations indicate that different neutrino species are transformed into each other. We show that this phenomenon can be explained if the neutrinos have non-vanishing masses, contrary to earlier assumptions.

7.1 W and Z Bosons

As described in the introductory chapter (Chap. 1), the weak interaction is responsible for the decay of neutrons: $n \rightarrow p + e^- + \bar{\nu}$. Neutron decay is a consequence of the decay of a d quark (which, however, decays only if the process is allowed by energy conservation): $d \rightarrow u + e^- + \bar{\nu}$.

As in the cases of the electromagnetic and strong interactions, this process is generated by the exchange of a particle with spin $\hbar$, the carrier of the weak interaction: the $W^{\pm}$ boson, which carries a positive or a negative electric charge (see Fig. 7.1); the W^- boson is the anti-particle of the W^+ boson.

In the case of the strong interaction there exist three "colors" q_i^s ($i = 1, 2, 3$), and a quark of a given color can be represented by a three-component vector in color

U. Ellwanger, *From the Universe to the Elementary Particles*,
Undergraduate Lecture Notes in Physics, DOI: 10.1007/978-3-642-24375-2_7,
© Springer-Verlag Berlin Heidelberg 2012

Fig. 7.1 Decay of a d quark
by the weak interaction

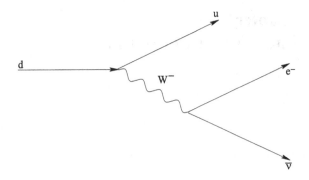

space. A change of its color (by the emission or absorption of a gluon) corresponds to a rotation of this vector. However, all other properties of a quark, such as its mass and its electric charge, are independent of the direction of this vector in color space.

In the case of the weak interaction there exists a quantity corresponding to the three colors, which is called, for historic reasons, *weak isospin*. Weak isospin can assume only two different values, denoted by "up" and "down". In contrast to color, the physical properties of particles with different values of weak isospin are not the same: the electric charge of a particle with isospin "up" is always larger by $+e$ than the charge of the corresponding particle with isospin "down", and their masses differ as well.

Quarks *and* leptons carry weak isospin: the quarks u and d have isospin "up" and isospin "down", respectively. Likewise, the neutrino carries isospin "up" and the electron isospin "down".

As in the case of the strong interaction, where the emission or absorption of a gluon changes the color of a quark, the isospin of a particle is changed by the emission or absorption of a $W^{\pm}$ boson. This explains the first part of Fig. 7.1, where a d quark is transformed into a u quark by the emission of a W^{-} boson. The second part of Fig. 7.1, the decay of a W^{-} boson into an electron and a(n) (anti)neutrino, is obtained as follows: take a vertex where a neutrino transforms into an electron by the absorption of a W^{-} boson, and reverse the chronological direction of the neutrino line (see the transformation of Fig. 5.4 into Fig. 5.9 in Chap. 5). This is why the neutrino in the final state has to be replaced by its antiparticle $\bar{v}$.

In fact, u and d quarks are not the only ones with isospin "up" and "down": also c and s quarks, as well as t and b quarks, form pairs with isospin "up" and "down", and can be represented as two-component vectors in isospin space:

$$\begin{pmatrix} u \\ d \end{pmatrix} \begin{pmatrix} c \\ s \end{pmatrix} \begin{pmatrix} t \\ b \end{pmatrix} \tag{7.1}$$

These three pairs of quarks are also called the three quark families. In parallel, there exist three families of lepton pairs. Up to now we have introduced only the electron and its neutrino, since the electron is the only stable charged lepton. However, we know of three charged leptons, the electron e^{-}, the muon μ^{-}, and the τ^{-} lepton.

Fig. 7.2 Possible emissions of W⁻ and W⁺ bosons by quarks

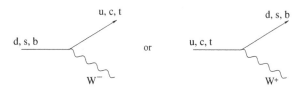

Their masses are

$$m_{e^-} \simeq 0.511\,\mathrm{MeV}/c^2,$$
$$m_{\mu^-} \simeq 106\,\mathrm{MeV}/c^2,$$
$$m_{\tau^-} \simeq 1.78\,\mathrm{GeV}/c^2. \tag{7.2}$$

For each charged lepton there exists a corresponding neutrino. (The 1988 Nobel prize was awarded to L.M. Ledermann, M. Schwartz, and J. Steinberger for the discovery of the muon neutrino ν_μ, and the 1995 Nobel prize went to M.L. Perl for the discovery of the τ lepton.)

The neutrino masses are very small; the electron neutrino is lighter than about $2\,\mathrm{eV}/c^2$. Indeed, it has only been known for a few years that their masses are not exactly zero: we have observed processes where a neutrino of a given family transforms into a neutrino of another family. This is possible only if they have slightly different masses (see Sect. 7.5); however, at present we have very little information (apart from upper bounds) on the absolute values of these masses

The three pairs of leptons with isospin "up" and "down" can be represented as follows:

$$\begin{pmatrix} \nu_e \\ e^- \end{pmatrix} \quad \begin{pmatrix} \nu_\mu \\ \mu^- \end{pmatrix} \quad \begin{pmatrix} \nu_\tau \\ \tau^- \end{pmatrix} \tag{7.3}$$

The emission or absorption of a $W^\pm$ boson by a lepton transforms it always into the corresponding lepton of the same family. This is not always true in the case of quarks: a quark of a given family can transform into a quark of another family, which leads to a large number of possible vertices, as shown in Fig. 7.2. The relative strengths of the couplings of quarks of different families to $W^\pm$ bosons are given by the elements of the so-called *Cabibbo–Kobayashi–Maskawa matrix*.

Correspondingly, a W^- boson can decay as in Fig. 7.3 in $3 \times 3 = 9$ different ways into a quark and an antiquark (if the process is allowed by energy conservation).

In addition, leptonic decays of W^- and W^+ bosons are possible. The vertices in Fig. 7.2 allow for decays of heavy quarks (s, c, b, t) into lighter quarks, accompanied by the decay of a $W^\pm$ boson into a quark–antiquark or lepton pair, as shown in Fig. 7.4 (or u, c, t → d, s, b + W^+, etc.), and decays of heavy leptons (μ, τ) into their neutrinos, accompanied by the decay of a W^- boson into a quark–antiquark or lepton pair, as illustrated in Fig. 7.5.

Fig. 7.3 Possible decays of a
W⁻ boson into quarks

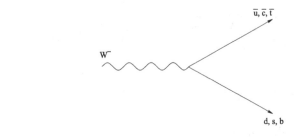

Fig. 7.4 Possible decays of
the d, s, or b quark by the
weak interaction

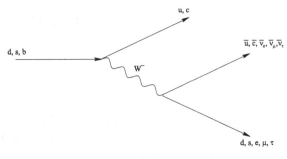

Fig. 7.5 Possible decays of a
τ lepton by the weak
interaction

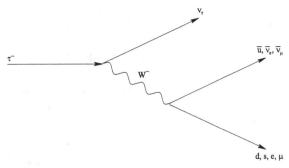

All these processes take place if and only if they are allowed by energy conservation. In decays such as in Figs. 7.4 and 7.5, we will find only particles of total mass smaller than the mass of the decaying quark or lepton in the final state: b quarks can never decay into t or t̄ quarks (or into other b quarks), and τ⁻ leptons only either "leptonically" into electrons or muons (and their neutrinos), or "hadronically' into a ū quark and a d or s quark, which will subsequently form hadrons. Muons can decay only into electrons and the corresponding neutrinos.

Finally, only the light quarks u and d and the electron remain stable. (The d quark remains stable inside protons and neutrons inside nuclei if the final state after a decay d → u + e⁻ + ν̄ would have a larger mass than the initial state.)

An essential difference between the weak interaction and the electromagnetic and strong interactions is that the W± bosons are massive:

$$M_W \simeq 80.4 \, \text{GeV}/c^2 \, . \tag{7.4}$$

From this fact we can deduce the "weakness" of the weak interaction. Historically, the mass of the $W^{\pm}$ bosons was postulated in order to explain this "weakness"; the mass of the $W^{\pm}$ bosons was confirmed only in 1983 in the UA1 detector experiments at the proton–antiproton storage ring SPS at CERN, Geneva; the 1984 Nobel prize was awarded to Rubbia und van der Meer for the discovery of the $W^{\pm}$ bosons.

First we consider the weak charge or, better, the weak fine structure constant α_w; it is even somewhat larger than the electromagnetic fine structure constant α. Since the weak interaction affects the electric properties of the particles (it changes their electric charge), the weak coupling and the electric one are not independent. Their ratio is given by a *weak mixing angle* or *Weinberg angle* θ_w (details will be discussed in Sect. 9.3):

$$\alpha_w = \alpha / \sin^2 \theta_w \simeq 3.4 \times 10^{-2}, \tag{7.5}$$

where the square of the sinus of the weak mixing angle is

$$\sin^2 \theta_w \simeq 0.22. \tag{7.6}$$

How is it thus possible that the weak interaction is weaker than the electromagnetic interaction? The reason can be found in the calculation of probabilities of processes induced by the exchange of $W^{\pm}$ bosons. One of the various factors entering the calculation of the probability $P(\theta)$ in the case of electron–electron scattering in Chap. 5 was the photon propagator $\mathcal{P}$. It depends on the difference between the actual energy E^{ph} of the photon and its classical value $E_{\text{cl}}^{\text{ph}}(\vec{p}^{\text{ph}}) = c \left| \vec{p}^{\text{ph}} \right|$ as a function of the momentum:

$$\mathcal{P}(E^{\text{ph}}, \vec{p}^{\text{ph}}) = \frac{-\hbar}{(E^{\text{ph}})^2 - [E_{\text{cl}}^{\text{ph}}(\vec{p}^{\text{ph}})]^2}. \tag{7.7}$$

The larger this difference, the smaller the propagator and, accordingly, the smaller the probability of the corresponding process.

In the calculation of the probability of a process induced by the exchange of W^- bosons as in Fig. 7.1 there appears the propagator (squared) of the W^- boson. The expression for this propagator is also given by

$$\mathcal{P}(E^W, \vec{p}^W) = \frac{-\hbar}{(E^W)^2 - [E_{\text{cl}}^W(\vec{p}^W)]^2}, \tag{7.8}$$

where the formula (3.22) has to be used for $[E_{\text{cl}}^W(\vec{p}^W)]^2$:

$$[E_{\text{cl}}^W(\vec{p}^W)]^2 = M_W^2 c^4 + (\vec{p}^W)^2 c^2. \tag{7.9}$$

The energy E^W and the momentum $\vec{p}^W$ of the W^- boson are determined via energy and momentum conservation by the energies and the momenta of the d and u quarks before and after the emission of the W^- boson.

Now it is easy to see that the term $M_W^2 c^4$ in the denominator of the propagator (7.8) dominates. To this end it suffices to consider the orders of magnitude of the different terms in the W^- propagator: the d and u quarks are inside a neutron (before the emission) or inside a proton (after the emission). Their energies and also their momenta multiplied by c are all of the order $m_p c^2 \sim 1$ GeV. Consequently both the actual energy E^W of the W^- boson and $|\vec{p}^{\,W}|\,c$ are of the order $m_p c^2$ and negligible compared to $M_W c^2$. Thus the propagator simplifies to

$$\mathcal{P}(E^W, \vec{p}^{\,W}) \sim \frac{\hbar}{M_W^2 c^4}. \tag{7.10}$$

All other dimensionful quantities, such as the momenta and energies of the leptons, are also of the order of the proton mass (multiplied by corresponding powers of c). The final result for the probability—proportional to the square of the W^- propagator—is thus necessarily proportional to $m_p^4 / M_W^4 \sim 10^{-7}$.

This consideration is valid for all processes induced by the exchange of $W^\pm$ bosons, as long as they take place at energies $\ll M_W c^2$. Accordingly these processes are relatively unlikely, i.e., slow, and the lifetimes τ of particles decaying by the weak interaction are relatively long. For the neutron, the muon, and the τ lepton they are

$$\tau_n \simeq 885.7\,\text{s}, \qquad \tau_\mu \sim 2.2 \times 10^{-6}\,\text{s}, \qquad \tau_\tau \sim 2.9 \times 10^{-13}\,\text{s}. \tag{7.11}$$

Although these lifetimes differ by many orders of magnitude due to the different particle masses, they are all relatively large compared to the lifetime of particles decaying by the strong interaction, for example, $\tau_\Delta \sim 5 \times 10^{-24}$ s for the Δ baryons in the previous chapter.

In the framework of the weak interactions there exists an additional massive boson, the neutral Z^0 boson, whose exchange also generates "weak" processes. The mass of the Z^0 boson was predicted in terms of the mass of the $W^\pm$ bosons and the weak mixing angle θ_w, and this value was confirmed after its discovery:

$$M_Z \simeq M_W / \cos\theta_w \simeq 91.2\,\text{GeV}/c^2. \tag{7.12}$$

As in the case of the photon, the emission or absorption of a Z^0 boson does not change the nature of the corresponding particle. However, neutrinos, which do not couple to the photon, can emit or absorb Z^0 bosons, since *all* particles with weak isospin couple to Z^0 bosons.

7.2 Parity Violation

In Sect. 5.4 we stated that quarks and leptons are fermions with spin (internal angular momentum) $\hbar/2$. Also, the spin of fermions can be visualized most easily as circularly polarized waves, whose polarization vector rotates in the plane perpendicular to the direction of propagation of the wave as in Fig. 5.12 or in Fig. 5.13. A priori, the wave equation for fermions allows for right- and left-handed polarized waves. The corresponding particles are denoted as right- or left-handed fermions.

Astonishingly, these fermions can possess different properties in the form of different charges, or different couplings to bosons whose exchange is responsible for interactions.

This shows again that the idea of a rotating sphere for an elementary particles with spin cannot be correct: a sphere rotating around an axis can always be tilted by $180°$ such that it rotates around the same axis in the opposite sense, without changing its charge. However, in the case of fermions we have to interpret states with spins parallel or antiparallel to the direction of propagation as different particle species in general.

The only interaction that distinguishes right- and left-handed fermions is the weak interaction: W bosons (W^+ and W^-) couple exclusively to left-handed, not to right-handed fermions. This holds for all quarks and leptons (charged or neutral). Only in the case of antiquarks and antileptons is the rule reversed: W bosons couple exclusively to right-handed antifermions. All processes shown in Figs. 7.1–7.5 occur only for the corresponding left-handed quarks or leptons and right-handed antiquarks or antileptons.

Likewise, the representation of quarks and leptons as two-component vectors with isospin "up" and "down" in (7.1) and (7.3) is valid only for left-handed particles; the right-handed quarks and leptons carry no weak isospin, and we have no evidence for right-handed neutrinos up to now.

This behavior of the weak interaction violates a symmetry denoted as *parity*. A *parity transformation P* is defined in which the directions of all three (x-, y-, and z-) axes are reversed. It is easy to see that the Klein–Gordon equation (4.1) is invariant under these transformations of the variables x, y, and z, since the corresponding derivatives appear only as squares. If all fundamental equations and quantities were invariant under parity transformations, this symmetry would manifest itself as a symmetry of all observable processes.

The world after a parity transformation corresponds to a world seen through a mirror: a parity transformation can be decomposed into a rotation by $180°$ around the z-axis (which reverses the x- and y-axes), and a last reflection in the x–y plane, reversing the z-axis. Since rotations are always symmetries of fundamental equations as well as of observable processes (see (Chap. 9), the only questionable operation remains the reflection in the x–y (or any other) plane. Therefore a parity transformation is often identified with a reflection.

What becomes of right-handed (or left-handed) fermions after a parity transformation? As sketched in Fig. 7.6 it is easy to see that, after a parity transformation, the handedness is reversed: now, the directions of all vectors are reversed. Initially, the sense of rotation of the polarization vector (corresponding to the angular momentum vector $\vec{L}$ in Fig. 5.11) remains the same, but finally the direction of flight denoted as $\vec{v}$ is reversed as well. Thus the handedness, given by the sense of rotation of the polarization vector along the direction of flight, is reversed.

Fig. 7.6 All vectors—including the direction of flight denoted as $\tilde{v}$—are reversed after a parity transformation P

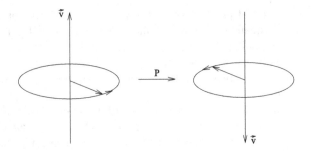

Hence, after a parity transformation, right-handed fermions become left-handed fermions, and vice versa. For this reason the world obtained after a parity transformation differs from ours: W bosons would couple to right-handed fermions. As a result of this property of the weak interactions, our world is *not* invariant under reflections!

Parity violation was observed first in 1956–1957 in the β decay of the cobalt-60 nucleus, in an experiment led by C.-S. Wu. The 1957 Nobel prize was awarded to T.-D. Lee and C.N. Yang for the theoretical interpretation of this discovery.

7.3 The Higgs Boson

Initially, the masses of the $W^{\pm}$ and Z^0 bosons were considered a serious problem: in quantum field theory, all carriers of interactions with spin $\hbar$ (such as the photon and gluons) should in principle be massless (see Sect. 9.3). This problem was solved by the discovery that masses of particles can be generated indirectly. This requires, however, the introduction of an additional boson with spin 0, denoted as the *Higgs boson* (after P.W. Higgs [12–14]; however, the corresponding mechanism was also developed independently by F. Englert and R. Brout [15] and G.S. Guralnik, C.R. Hagen, and T.W.B. Kibble [16, 17]. The 1999 Nobel prize was awarded to G.'t Hooft and M.J.G. Veltman for the proof of the mathematical consistency of such a theory).

In quantum field theory a particle always corresponds to a field: the photon to the electric and magnetic fields, and the gluon to a "gluonic" field (which exists only inside hadrons, however), etc. Likewise, the Higgs boson corresponds to a Higgs field. The electric and magnetic fields are vector fields (which point in a given direction), since the photon carries a spin $\hbar$. In contrast, the field corresponding to the Higgs boson is a scalar field.

To begin with, we consider the motion of a charged particle in a constant electric field: as a result of the Lorentz force (5.1) acting on the particle, its energy and its momentum are modified compared to the motion in the absence of an electric field. Clearly, these modifications are proportional to the electric charge of the particle, since the Lorentz force is proportional to this charge.

What would be the effect of a constant Higgs field on the motion of a particle? Now the momentum remains unchanged, just the energy increases by a (time-independent)

Fig. 7.7 Vertex of a particle
p and a Higgs boson

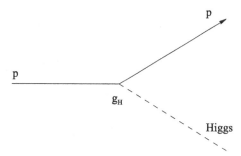

amount proportional to the value H of the Higgs field, and to a coupling constant g_H (a kind of charge) of the particle to the Higgs field:

$$\Delta E = g_H H \tag{7.13}$$

The coupling constant g_H is related, like the coupling constant g in (5.26), to a particle—Higgs vertex (see Fig. 7.7); its numerical value depends on the particle species p.

Now we assume that the particle is massless in the absence of a Higgs field. Then its energy is given by $E^2 = \vec{p}^2 c^2$. In the presence of a Higgs field, this expression has to be replaced by

$$E^2 = \vec{p}^2 c^2 + g_H^2 H^2. \tag{7.14}$$

This formula can be compared to (3.22) for the energy of a massive particle: the corresponding expressions are identical if we identify $m^2 c^4$ with $g_H^2 H^2$. In fact, the particle in a Higgs field behaves in all respects like a particle with mass

$$m = g_H H/c^2. \tag{7.15}$$

In this way we can "generate" a mass for each particle, if there exists everywhere a constant Higgs field H.

Why do we not find everywhere a constant electric field $\vec{E}$? The reason is that this would cost energy. This energy is denoted as potential energy E_{pot}, and is proportional to the square $\left|\vec{E}\right|^2$ of the electric field. (This is a consequence of the field equations of electrodynamics.) Similar to a point particle, a field assumes normally a configuration of lowest possible energy, and the stable configuration of lowest possible energy is $\vec{E} = 0$.

This argument can justify an everywhere-constant Higgs field—under the condition, however, that the dependence of the potential energy E_{pot} on the Higgs field H is different from its dependence on the electric field $\vec{E}$. In fact, the field equations for a scalar field allow an arbitrary dependence on the corresponding field; we can assume, e.g., that

Fig. 7.8 Schematic representation of the dependence of the potential energy on the Higgs field H corresponding to (7.16)

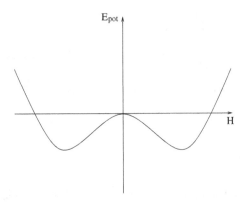

$$E_{\mathrm{pot}}(H) = \frac{1}{(\hbar c)^3}\left(-\frac{\mu^2}{2}H^2 + \frac{\lambda_{\mathrm{H}}^2}{4}H^4\right), \tag{7.16}$$

which is of the form shown in Fig. 7.8.

(In principle the Higgs field is a two-component vector in isospin space, like the quarks and leptons in (7.1) and (7.3). Here we consider only the neutral component H, which, owing to the Higgs–Kibble mechanism not discussed here, is the only one corresponding to a physical particle.)

The coefficients in (7.16) are chosen such that $E_{\mathrm{pot}}(H)$ has the dimension of an energy density (energy per cubic meter), μ is measured in GeV, and λ_{H} is a dimensionless constant. The minimum of this function of H is at $H = \pm\mu/\lambda_{\mathrm{H}}$, and not at $H = 0$. The Higgs field H will assume one of the values (e.g., the positive one) with minimal potential energy everywhere in the Universe; any other value would correspond to an unstable configuration.

A possible explanation for the mass of the $\mathrm{W}^{\pm}$ and Z^0 bosons—in fact the only coherent one in quantum field theory—is thus the existence of a Higgs field, a dependence of the potential energy on H as in (7.16), and a weak charge of the Higgs boson: this implies a coupling g_{H} of the $\mathrm{W}^{\pm}$ to the Higgs boson given by $g_{\mathrm{H}} = g_{\mathrm{w}}/2$. ($g_{\mathrm{w}}$ is related to α_{w} in (7.5) by $\alpha_{\mathrm{w}} = g_{\mathrm{w}}^2/(4\pi)$, i.e., we have chosen g_{w} dimensionless here. The coupling of the Z^0 boson to the Higgs boson is given by $g_{\mathrm{w}}/2\cos\theta$.) Then we obtain from (7.15)

$$M_{\mathrm{W}} = \frac{g_{\mathrm{w}}}{2c^2}H, \tag{7.17}$$

which, using $g_{\mathrm{w}} \sim 0.65$, gives the value (7.4) for M_{W} if the value of the Higgs field H is

$$H \simeq 248\,\mathrm{GeV}. \tag{7.18}$$

For the photon and the gluons we obtain no masses generated by the Higgs field, since the couplings of the photon and the gluons to the Higgs field vanish since the Higgs boson carries neither an electric nor a strong charge (i.e., color).

The masses of the quarks and the leptons can also be explained by couplings to the Higgs field. These couplings are called Yukawa couplings. (H. Yukawa, 1949 Nobel prize, was the first to introduce couplings of fermions with spin $\hbar/2$ to scalar fields in 1935 (at that time between protons, neutrons, and pions) in order to describe the strong interactions between baryons in terms of an exchange of pions.) These couplings are denoted by λ_i, where the index i corresponds to the quark or lepton. Hence we have from (7.15)

$$m_e = \lambda_e H/c^2 \quad \ldots \quad m_{\text{top}} = \lambda_{\text{top}} H/c^2. \tag{7.19}$$

Unfortunately these formulas do not allow us to compute the masses of the quarks and leptons: whereas we know the value of the Higgs field H from (7.18), we cannot predict the numerical values of the Yukawa couplings λ_i. All we can do is to determine the Yukawa couplings from the known quark and lepton masses,

$$\lambda_e \simeq 2 \times 10^{-6} \quad \ldots \quad \lambda_{\text{top}} \simeq 0.7. \tag{7.20}$$

To date we know of no satisfactory explanation for the enormous differences between the Yukawa couplings.

We should add that there exists another contribution to the masses of quarks: inside the hadrons there exists a gluon field leading to a similar contribution to the energy of quarks as the Higgs field in (7.14). This contribution of the gluon field to the quark masses amounts to about 300 MeV/c^2; it explains practically all of the masses of the u and d quarks, to which the Higgs field contributes just a few MeV/c^2.

Although the explanation of the masses of all particles (in particular the $W^\pm$ and Z^0 bosons) predicts the presence of a Higgs boson, it does not allow a prediction of the mass M_H of this particle itself: on the one hand, in quantum field theory the mass squared of a scalar particle is proportional to the second derivative of the potential energy $E_{\text{pot}}(H)$ at the minimum, which gives, according to (7.16), $M_H = \mu/c^2$. On the other hand, the known value of $H = \mu/\lambda_H$ does not allow one to determine the parameter μ independently of the unknown parameter λ_H.

Hence we can only try to produce Higgs bosons in particle accelerators, and to measure their mass subsequently. However, in accelerators we can only produce particles whose mass is smaller than the available total energy. Up to November 2000 the accelerator LEP at CERN (Geneva) allowed a search for Higgs bosons with masses $M_H \leq 114\,\text{GeV}/c^2$, but no signal of a Higgs boson was discovered (see Chap. 8). Hence its mass is either larger than this value, or the situation is more complicated than assumed. (For instance, there could exist several Higgs bosons with smaller couplings, which would reduce their production rates and complicate their detection correspondingly.)

In 2009 the (LHC) Large Hadron Collider, a new more energetic accelerator, was put into operation at CERN. Its energy should suffice to clarify the situation in the

Higgs sector of the weak interaction. Details of the search for the Higgs boson at the LHC will be discussed in Chap. 8.

Finally we want to discuss a possible conflict between the expression (7.16) for the potential energy and the evolution of the Universe discussed in Chap. 2: the latest determinations of the cosmological constant Λ (or the "dark energy") found a value on the order of a few $10^{-10} \, \text{kg s}^{-2} \, \text{m}^{-1}$, given in (2.18). The potential energy introduced in (7.16) contributes to the cosmological constant; the corresponding contribution has to be evaluated for a Higgs field H at the minimum of the potential energy. Assuming $\lambda_H \sim 1$, this contribution can be estimated, and we obtain

$$E_{\text{pot}}(H = 248 \, \text{GeV}) \simeq -10^{44} \, \text{kg s}^{-2} \, \text{m}^{-1}. \tag{7.21}$$

This contribution is of the wrong sign, but first and foremost its absolute value is 54 orders of magnitude larger than the measured value!

It is true, in fact, that the expression (7.16) for the potential energy can be replaced by

$$E_{\text{pot}}(H) = \frac{1}{(\hbar c)^3}\left(-\frac{\mu^2}{2}H^2 + \frac{\lambda_H^2}{4}H^4\right) + C. \tag{7.22}$$

The constant C does not change the value of H at the minimum of $E_{\text{pot}}(H)$, and the non-vanishing of H at the minimum is the only important property of this expression for the potential energy. In principle we can find a value for C such that the new value of $E_{\text{pot}}(H = 248 \, \text{GeV})$ agrees with (2.18). However, nobody has a reasonable idea about the origin of a constant C such that C assumes precisely this value—to a precision of 54 decimals! (Moreover, the situation is complicated by the fact that there exist additional quantum contributions to the potential energy that have nothing to do with the Higgs field, but which are at least of the same order and must also be compensated by a corresponding value of C.)

This "cosmological constant problem", one of the open questions in cosmology (see Sect. 2.5), preoccupies both cosmologists and theoretical particle physicists.

Assuming that the constant C is such that the value of the minimum of the function $E_{\text{pot}}(H)$ is essentially $E_{\text{pot}}(H_{\text{min}}) = 0$ allows an interesting speculation concerning the origin of inflation discussed in Sect. 2.4: inflation corresponds to an extremely rapid exponential expansion of the Universe (see 2.20) and explains the uniform distribution of galaxies and the cosmic background radiation in today's Universe. Inflation takes place if the parameter Λ ($=E_{\text{pot}}$) in the Friedmann–Robertson–Walker equations (2.6) and (2.7) is much larger than $\varrho(t)c^2$ and $p(t)$, but an inflationary epoch must also have come to an end. This would be the case if E_{pot} was temporarily very large and decreased only later to its value today compatible with (2.18). Since E_{pot} depends on all fields present in the Universe, E_{pot} can vary if at least one of these fields varies in the course of time.

As mentioned above, fields assume normally the value minimizing the potential energy. At the beginning of the Universe, the values of fields can have been different, however; we can assume, e.g., that at the beginning of the Universe the value of the

Fig. 7.9 Schematic representation of the origin of a time-dependent field H and hence a time-dependent potential energy $E_{\text{pot}}(H)$

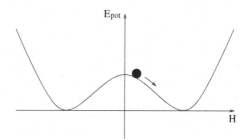

Fig. 7.10 Schematic representation of a time dependence of a field Φ allowing to satisfy the inequality (7.23)

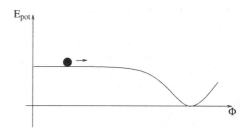

Higgs field H was close to 0. Then the value of H will change as sketched in Fig. 7.9: H will "roll" into the minimum of the function $E_{\text{pot}}(H)$.

In doing so, the potential energy decreases until it nearly vanishes at the end (owing to the assumed value of the constant C); in principle this is the behavior desired in a model for inflation.

Regrettably it turns out that, for a potential energy as in Fig. 7.9, the length of time of the inflationary epoch is not long enough: we mentioned in Sect. 2.4 below (2.20) that the interval Δt of the inflationary epoch should satisfy the inequality

$$\sqrt{\frac{\kappa \Lambda}{3}} \Delta t \gtrsim 60. \tag{7.23}$$

Only then can a region with diameter Δd, within which the original gas was distributed homogeneously, inflate to a volume larger than the Universe known today. (In (7.23) we have to replace Λ by $E_{\text{pot}}(H)$.) However, in Fig. 7.9 the Higgs field drops too fast into the minimum, so that the inequality (7.23) is not satisfied.

In order to satisfy the inequality (7.23), the potential energy should depend on a field Φ such that the duration of stay of the field at a large value of $E_{\text{pot}}(\Phi)$ (playing the role of Λ) is extended. This would be the case, e.g., for a shape of $E_{\text{pot}}(\Phi)$ as in Fig. 7.10.

Regrettably a field Φ with a potential energy of the form in Fig. 7.10 (with parameters such that the inequality (7.23) is satisfied) cannot be identified as the Higgs field of the weak interaction. Hence many cosmologists believe that additional fields exist, whose potential energy is of the form in Fig. 7.10 and which could be responsible for an inflationary epoch in the early Universe.

7.4 CP Violation

Besides the parity transformation P discussed in Sect. 7.2, there exists another transformation denoted by *charge conjugation* C: charge conjugation corresponds to all particles being replaced by their antiparticles (with opposite charges) and vice versa. At first all interactions seem to be invariant under this replacement but, taking a closer look, we find that this is not true for the weak interaction: after a charge conjugation, a left-handed fermion, with couplings to W bosons, becomes a left-handed antifermion, which does *not* couple to W bosons. Thus charge conjugation is not a symmetry of the weak interaction.

However, we can consider the product of two transformations, charge conjugation C and parity transformation P; this product is called a CP transformation.

After a CP transformation a left-handed fermion becomes a right-handed antifermion coupling to W bosons just like a left-handed fermion. Thus, for a long time it was believed that all fundamental interactions are invariant under a CP transformation.

This idea turned out to be wrong, as we learned from the study of decays of K^0 mesons: the two existing K^0 mesons are electrically neutral mesons of spin 0, corresponding to different superpositions of the quark states $d\bar{s}$ and $\bar{d}s$. They have to be described by fields K_1 and K_2, which are either invariant under a CP transformation ($K_1 \rightarrow K_1$), or turn into their negative ($K_2 \rightarrow -K_2$). The fields K_1 are called even, and the fields K_2 odd under CP; we can associate CP charges $+1$ and -1 to K_1 and K_2, respectively.

If all fundamental interactions were invariant under CP transformations, the fields K_1 could decay only into a state equally even under CP, e.g., into two pions. This decay is relatively fast, leading to a short lifetime of K_1. (K_1 is also denoted as K_{short}.) K_2 should only decay into a state equally odd under CP, e.g., into three pions. This decay is relatively slow; accordingly the mean lifetime of K_2 is relatively long. (K_2 is also denoted as K_{long}.)

Astonishingly it was observed that, albeit very rarely, K_{long} can also decay into two pions. Since this final state transforms differently under CP transformations than the initial state, an interaction *not* invariant under CP transformations must have occurred; we talk about an observed *CP violation*. The 1980 Nobel prize was awarded to J.W. Cronin and V.L. Fitch for the discovery of this phenomenon in 1964.

The only parameters capable of violating symmetry under CP transformations within the hitherto described theory of weak interactions are the Yukawa couplings in (7.20). We cannot discuss the details here, and mention only that symmetry violation under CP transformations is related to the fact that these Yukawa couplings can be complex (see Chap. 9), implying that also the quark masses and finally the couplings of quarks to W bosons (the Cabibbo–Kobayashi–Maskawa matrix elements) can be complex quantities. Up to now it looks as if this origin of CP violation can describe the behavior of the K mesons (and others, such as the B mesons, which are presently under study).

Symmetry violation by the fundamental interactions under CP transformations also has important cosmological consequences: at first sight we would expect that, according to the theory of the Big Bang, we would be left with as many particles as antiparticles of every species, which would tend to annihilate practically completely after the numerous scattering and decay processes of elementary particles at the beginning of the Universe (during the very hot and compressed epoch). This does not agree with the observation that there exist practically no antiparticles in the present Universe. One can show that a particle–antiparticle disequilibrium can be generated spontaneously only if at least one of the fundamental interactions is *not* invariant under CP transformations.

Hence the circle closes: the observed CP violation can be responsible for the particle–antiparticle disequilibrium in today's Universe; without this disequilibrium all particles would have annihilated with antiparticles during the Big Bang, making our existence impossible. At present researchers are trying to compute, taking CP violation into consideration, the (measured) number of remaining particles after the scattering and decay processes of elementary particles at the beginning of the Universe. It seems, however, that the particle–antiparticle asymmetry can be understood only if one assumes the existence of more elementary particles than those known at present.

7.5 Neutrino Oscillations

We have seen in Sect. 7.1 that each charged lepton (electron, muon, or τ) is accompanied by its own neutrino. Neutrinos are generated in decays induced by the weak interaction. The energy and momentum that they carry away led to their existence being postulated. Comparing their energy and their momenta to the relativistic formula (3.22), we can obtain information on their masses. However, this way we have only found up to now that these masses must be very small compared to their energies and momenta (multiplied by the corresponding powers of c); the upper bounds are as follows:

$$m_{\nu_e} \lesssim 2\,\text{eV}, \qquad m_{\nu_\mu} \lesssim 190\,\text{keV}, \qquad m_{\nu_\tau} \lesssim 18\,\text{MeV} . \qquad (7.24)$$

Therefore it was believed for a long time that neutrino masses vanish exactly.

Neutrinos are very difficult to detect. All the other elementary particles, such as quarks (inside hadrons) or charged leptons, scatter either off atomic nuclei or off the electrons of atomic shells via the strong or the electromagnetic interaction. Neutrinos are the only elementary particles sensitive only to the weak interaction, and the probabilities of such processes are very small. Therefore a single neutrino can pass through, with a very high probability, enormous amounts of matter (for instance the entire Earth) without a scattering process. Millions upon millions of neutrinos can cross a body with only very few of them reacting with single atoms. In order to detect neutrinos, we have to observe carefully literally thousands of tons of iron or of water with the help of detectors in order to discover the few processes induced by neutrinos; moreover, these processes have to be distinguished from other processes that are due to natural radioactivity or cosmic radiation.

Neutrinos can be produced artificially in the β decay of neutrons produced in nuclear reactors (see Sect. 1.3), or in the decays of mesons such as pions and K mesons (produced in accelerator experiments, see Chap. 8) into electrons or muons and the corresponding neutrinos.

However, "natural" neutrino sources exist as well: our atmosphere is permanently being bombarded by very energetic cosmic radiation, which consists of about 90% protons but also α particles (helium nuclei), electrons, and photons. (The 1936 Nobel prize was awarded to V. Hess for the discovery of cosmic radiation.) When these particles hit the atmosphere, they create first avalanches of photons, electrons, and hadrons (mainly pions). Subsequently, these decay into so-called atmospheric neutrinos, which reach the surface of the Earth.

In addition, in the interior of the Sun nuclear reactions take place, which keep the sun shining and produce so-called solar neutrinos. Finally we have to expect that neutrinos were produced during the Big Bang, have not been absorbed since, and transit the Universe in a similar way to the cosmic background radiation. Also, astrophysical processes such as supernova explosions contribute to the production of cosmic neutrinos.

A surprising discovery during the last few years was that different species of neutrinos ν_e, ν_μ and ν_τ can transform into each other, i.e., oscillate. The first hint of these so-called neutrino oscillations originated from attempts to detect solar neutrinos.

We know quite precisely which nuclear reactions take place inside the Sun, and with which abundance they produce electron neutrinos of a given energy. Hence we know with which rate they should be detected on Earth. However, the measured detection rate is about only half of that expected; instead we find too many muon neutrinos with the corresponding energies.

In the case of atmospheric neutrinos, we find another anomaly: as we can verify in accelerator experiments, pions generated by the cosmic radiation decay on average into about twice as many muons and their neutrinos as into electrons and their neutrinos. However, the measured ratio of the ν_μ to ν_e rates is smaller. This means that muon neutrinos have disappeared. Since the number of electron neutrinos has not increased, we assume that the muon neutrinos have mutated into τ neutrinos. This interpretation agrees with the behavior of muon neutrinos produced in accelerator experiments.

Hence all three neutrino species seem to mutate into each other. Interestingly enough, a theoretical description of this phenomenon is possible only if they are *not* exactly massless. This theoretical description employs again the equivalence of a beam of neutrinos and a wave solution of the Klein–Gordon equation discussed in Sect. 4.2.

In the following we will sketch the theoretical description of this phenomenon, confining ourselves to two neutrino species. These two kinds of neutrinos correspond to fields $\Psi_1(x, t)$ and $\Psi_2(x, t)$; we assume again that the two waves are directed along the x-axis, hence are independent of y and z.

If the neutrinos were massless, both fields $\Psi_1(x, t)$ and $\Psi_2(x, t)$ would satisfy the massless Klein–Gordon equation (4.1) or the simplified version (4.2). However, neutrino oscillations appear only if both fields $\Psi_1(x, t)$ and $\Psi_2(x, t)$ satisfy equations with mass terms that mix the fields, for example,

$$\left(\frac{\partial^2}{\partial t^2} - c^2 \frac{\partial^2}{\partial x^2} + \frac{m^2 c^4}{\hbar^2} \right) \Psi_1 + \frac{\Delta m^2 c^4}{2\hbar^2} \Psi_2 = 0,$$
$$\left(\frac{\partial^2}{\partial t^2} - c^2 \frac{\partial^2}{\partial x^2} + \frac{m^2 c^4}{\hbar^2} \right) \Psi_2 + \frac{\Delta m^2 c^4}{2\hbar^2} \Psi_1 = 0. \tag{7.25}$$

Here we are dealing with a system of two coupled partial differential equations. This system can be decoupled if we introduce the fields

$$\Psi_+ = \Psi_1 + \Psi_2, \qquad \Psi_- = \Psi_1 - \Psi_2, \tag{7.26}$$

and consider the sum and the difference of the equations (7.25) above. Then we find that the fields Ψ_+ and Ψ_- satisfy the massive Klein–Gordon equations with $m_+^2 = m^2 + \Delta m^2/2$, $m_-^2 = m^2 - \Delta m^2/2$, i.e.,

$$\Delta m^2 = m_+^2 - m_-^2. \tag{7.27}$$

However, the fields Ψ_1 and Ψ_2 are the ones corresponding to a given species of neutrinos (e.g., to the electron or muon neutrinos): in weak interaction processes, electron or muon neutrinos will always be produced together with the corresponding charged lepton. Likewise, at the moment of their detection in scattering processes via the exchange of $W^\pm$ bosons, they transform into the corresponding charged lepton—this way the neutrino species can be determined experimentally.

An interesting exact solution of the two equations (7.25) is given by the following expressions:

$$\Psi_1(x, t) = \cos(kx - \omega t) \cos\left(\frac{\Delta m^2 c^2}{4k\hbar^2} x \right),$$
$$\Psi_2(x, t) = \sin(kx - \omega t) \sin\left(\frac{\Delta m^2 c^2}{4k\hbar^2} x \right), \tag{7.28}$$

where ω satisfies

$$\omega^2 = k^2 c^2 + \frac{m^2 c^4}{\hbar^2} + \frac{\Delta m^4 c^6}{16k^2 \hbar^4}. \tag{7.29}$$

We recall that now $|\Psi_i(x, t)^2|$ corresponds to the probability of finding a neutrino of species i at the position x at the moment t. At a fixed position x, the time-dependent oscillations are usually so fast that measurements correspond to an average over time.

The two t-dependent functions $\cos^2(kx - \omega t)$ and $\sin^2(kx - \omega t)$ in $\Psi_i(x, t)^2$ (which both oscillate between 0 and 1) give the same average over the time t, denoted by a bar:

$$\overline{\cos^2(kx - \omega t)} = \overline{\sin^2(kx - \omega t)} = 1/2. \tag{7.30}$$

Hence we obtain for the solutions (7.28)

$$\overline{\Psi_1^2}(x) = \frac{1}{2}\cos^2\left(\frac{\Delta m^2 c^2}{4k\hbar^2}x\right), \qquad \overline{\Psi_2^2}(x) = \frac{1}{2}\sin^2\left(\frac{\Delta m^2 c^2}{4k\hbar^2}x\right). \tag{7.31}$$

We see that the sum of the two time-averaged probabilities does not depend on x:

$$\overline{\Psi_1^2}(x) + \overline{\Psi_2^2}(x) = \frac{1}{2}. \tag{7.32}$$

The factor $1/2$ is just a convention, which could be changed by a multiplication of the fields by an arbitrary constant. It is important, however, that according to (7.32) no particles get lost; the probability of detecting *any* particle species remains constant along the x-axis.

At the origin $x = 0$ of the x-axis we have $\overline{\Psi_1^2}(x) = 1/2$ and $\overline{\Psi_2^2}(x) = 0$. Therefore we can use this solution for the description of the situation where neutrinos of the kind Ψ_1 are produced at $x = 0$, and no neutrinos of the kind Ψ_2 are present to start with. For increasing x along the direction of flight, $\overline{\Psi_1^2}(x)$ decreases and $\overline{\Psi_2^2}(x)$ increases, until we have $\overline{\Psi_1^2}(x) = 0$ and $\overline{\Psi_2^2}(x) = 1/2$ for

$$x = \frac{L}{2} = \frac{2\pi k\hbar^2}{\Delta m^2 c^2}; \tag{7.33}$$

then, owing to the mixing term $\sim \Delta m^2$ in (7.25), all neutrinos of the kind Ψ_1 have mutated into neutrinos of the kind Ψ_2. Subsequently both probabilities oscillate as a function of x to and fro with a wavelength L, which can be obtained from (7.33). In practice, the mass terms in (7.29) are usually negligible; then k in (7.33) satisfies $k \simeq \omega/c \simeq E/(\hbar c)$, where we have used (4.8) and E denotes the energy of individual neutrinos. If we substitute this relation for k in (7.33), we obtain for the oscillation wavelength L

$$L \simeq \frac{4\pi E\hbar}{\Delta m^2 c^3}. \tag{7.34}$$

Clearly this description of neutrino oscillations implies that not all neutrinos can be massless: the two quantities m_+^2 and m_-^2 in (7.27) correspond, in principle, to measurable neutrino masses squared, and cannot vanish simultaneously if the oscillation wavelength L (and hence $1/\Delta m^2$) is less than infinity. On the other hand, the measurements of neutrino oscillation rates allow, as a matter of principle, only determinations of parameters such as Δm^2, not of the proper neutrino masses such as m_+ and m_-. (m_+ and m_- are the masses of particles corresponding to the fields Ψ_+ and Ψ_- with decoupled equations. In general, Ψ_+ and Ψ_- are superpositions of Ψ_1 and Ψ_2, see (7.26). Here the so-called mixing angle is $90°$, since we assumed the same mass term m^2 in both equations (7.25). If the value of m^2 were different in the two equations (7.25), the mixing angle would differ from $90°$.)

The oscillation wavelengths L can assume quite large values: the difference between the masses squared of the electron and muon neutrinos is compatible with $\Delta m^2 \sim 8 \times 10^{-5} \, \text{eV}^2/c^4$; if we use 1 MeV for E (a typical value for nuclear reactions), we find

$$L \simeq 30 \, \text{km}. \tag{7.35}$$

In nature we have to cope with three instead of two neutrino species. Then the mass terms in the three corresponding equations (7.25) can (and probably do) assume the form of 3×3 matrices. At present we have only fragmentary information about these mass terms. Numerous additional experiments are under way, under construction, or planned with the aim of determining the properties of neutrinos: experiments at nuclear reactors, at particle accelerators, and large-scale underwater (ANTARES and Baikal) and even under-ice (AMANDA in the Antarctic) experiments.

The various theories for the origin of these mass terms differ, among other things, in the presence of right-handed neutrinos, which would imply an even more complicated structure of neutrino mass terms. From the results of the experiments we hope to learn which of the different theories describes the way nature has chosen for the generation of neutrino masses.

Exercises

7.1. Give the complete list of quarks and leptons (lighter than an $\bar{s}$ quark) into which an $\bar{s}$ quark can decay by the weak interaction.

7.2. Use this result to deduce, taking a possible annihilation of the $u\bar{s}$ pair into account, the complete list of hadrons and leptons into which a K meson ($= u\bar{s}$) with a mass of $\sim 494 \, \text{MeV}/c^2$ can decay by the weak interaction (and additional strong interaction processes).

Chapter 8
The Production of Elementary Particles

Why do most of today's particle experiments require huge ring accelerators? The answer to this question is given and the layout of today's particle accelerators and detectors is outlined. As an example of the comparison of theory with experiment, we compute the production rate of hadrons in electron–positron collisions and compare it to the data. The fact that quarks occur with three different colors (of the strong interaction) is confirmed, as well as their electric charges and masses. Finally, we describe how the sought-after Higgs bosons can be produced, in particular at the Large Hadron Collider (the largest of today's particle accelerators), and which of the Higgs boson decays can be used for their detection.

8.1 Introduction to Accelerator Experiments

Experiments in particle physics serve the following purposes:

(a) To discover new particles, and/or to verify the existence of particles predicted in models.
(b) To measure their properties, i.e., their masses, charges, spins, and interactions (strong? weak? new interactions?) via their production and decay processes.

To this end in most cases two beams of stable particles—as energetic as possible—are used, e.g., electrons, positrons, protons, and antiprotons. Nowadays these particles are accelerated in ring accelerators, in which the two beams circulate in opposite directions. The two beams intersect at so-called intersection points, as shown in Fig. 8.1.

Most particles of a beam cross the intersection region without scattering off a particle of the other beam. These particles can be reused for later interactions—this possibility is one of the big advantages of ring accelerators. (For this reason they are also called storage rings.) If a particle scatters off a particle of the other beam, often numerous new particles are created. First, we have to distinguish the following kinds of scatterings.

U. Ellwanger, *From the Universe to the Elementary Particles*,
Undergraduate Lecture Notes in Physics, DOI: 10.1007/978-3-642-24375-2_8,
© Springer-Verlag Berlin Heidelberg 2012

Fig. 8.1 Two intersecting electron and positron beams

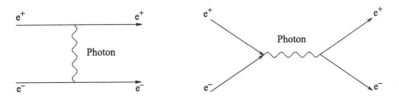

Fig. 8.2 Two diagrams contributing to elastic electron–positron scattering

Fig. 8.3 Creation of a particle–antiparticle pair in electron–positron annihilation

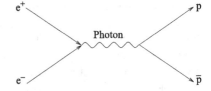

Elastic scattering: We talk about elastic scattering if the particles in the final state (after scattering) are the same as before scattering. However, their directions of flight have changed; see electron–electron scattering in Chap. 5. In the case of electron–positron scattering, elastic scattering is written as $e^+ + e^- \rightarrow e^+ + e^-$. Confining ourselves to the electromagnetic interaction, the two Feynman diagrams in Fig. 8.2 with two vertices contribute to this process. (In addition there exist diagrams arising from the weak interaction, where the photon is replaced by a Z boson.)

Inelastic scattering: Inelastic scattering corresponds to cases where the particles in the final state differ from the particles before scattering. Examples are processes of the kind $e^+ + e^- \rightarrow p + \bar{p}$, where p and $\bar{p}$ denote particles and antiparticles other than an electron and a positron. If the particles p and $\bar{p}$ carry electric charges, the Feynman diagram in Fig. 8.3 contributes to such a process.

Which particles p and $\bar{p}$ will be created this way? To answer this question we have to refer to a basic rule of quantum mechanics (and thus also of quantum field theory): all processes allowed by the laws of energy conservation, momentum conservation, and the conservation of electric charge (and possibly color or other charges) are possible, in principle. However, the different processes differ in their relative probabilities, i.e., their relative frequencies.

Before computing probabilities we have to verify under which condition energy is conserved. According to (3.24), the energy of a particle is given by $E =$

$\sqrt{m^2c^4 + |\vec{p}|^2 c^2}$. It follows from the law of conservation of energy that we have $E(e^+) + E(e^-) = E(p) + E(\bar{p})$, where $E(p)$ is the energy of the produced particle and $E(\bar{p})$ the energy of the antiparticle. From $E(p) > m_p c^2$, where m_p is the particle mass (equal to the mass of the antiparticle), we have

$$E(e^+) + E(e^-) = E = E(p) + E(\bar{p}) > 2m_p c^2. \tag{8.1}$$

Here E is the total energy of the process, which has to satisfy the inequality (8.1) in order that particle–antiparticle pairs of mass m_p can be produced.

First we have to resolve a little paradox: the energy depends on the momentum or the velocity, which, in turn, depends on the reference frame. The inequality (8.1) must be satisfied in any reference frame, but: in which reference frame can E be as small as possible such that particle–antiparticle pairs of a given mass m_p can be produced? This is the so-called center-of-mass frame, where the momenta of the incoming particles (and hence, owing to the conservation of the total momentum, the momenta of the outgoing particles) point in opposite directions. (The calculations in Sects. 5.2 and 5.3 have already been carried out in the center-of-mass frame.) Only in the center-of-mass frame can the the momenta of both produced particles be simultaneously minimal (practically equal to 0); hence both energies $E(p)$ and $E(\bar{p})$ can be minimal (practically equal to $m_p c^2$), and the total energy E must be hardly larger than $2m_p$. In other words, it is not enough that the total energy E is larger than $2m_p c^2$ in some reference frame; it is necessary and sufficient that this inequality holds in the center-of-mass frame.

Now we can explain why nowadays nearly all experiments are carried out at ring accelerators with two beams pointing in opposite directions, and not at accelerators with just one beam hitting a "fixed target". To this end we compute the total energy $E_{\text{ft(cm)}}$ in the center-of-mass frame of a fixed-target experiment, which differs greatly from the total energy $E_{\text{ft(lab)}}$ in the laboratory frame (the reference frame of the target at rest).

We assume that a beam of particles with momentum $\vec{p}_1$, mass m, and corresponding energy $E_1 = \sqrt{m^2c^4 + |\vec{p}_1|^2 c^2}$ hits a particle of the same mass at rest, and corresponding energy $E_2 = mc^2$. In the laboratory frame we have $E_{\text{ft(lab)}} = E_1 + E_2 = \sqrt{m^2c^4 + |\vec{p}_1|^2 c^2} + mc^2$, and the total momentum in the laboratory frame is $\vec{P}_{\text{ft(lab)}} = \vec{p}_1$. The simplest way to compute the total energy $E_{\text{ft(cm)}}$ in the center-of-mass frame is to use the fact that the combination $E^2 - c^2\vec{P}^2$ is the same in any reference frame (see (3.26)). In the center-of-mass frame the total momentum is $\vec{P} = 0$ by definition, hence we obtain

$$\begin{aligned} E_{\text{ft(cm)}}^2 &= E_{\text{ft(lab)}}^2 - |\vec{P}_{\text{ft(lab)}}|^2 c^2 \\ &= \left(\sqrt{\vec{p}_1^2 c^2 + m^2 c^4} + mc^2\right)^2 - \vec{p}_1^2 c^2 \\ &= 2\left(mc^2\sqrt{\vec{p}_1^2 c^2 + m^2 c^4} + mc^2\right). \end{aligned} \tag{8.2}$$

In today's experiments the momenta $|\vec{p}_1|$ are typically very large,

$$|\vec{p}_1| \gg mc. \tag{8.3}$$

Then we have

$$E^2_{\text{ft(cm)}} \simeq 2m|\vec{p}_1|c^3. \tag{8.4}$$

On the other hand, in the case of a ring accelerator with two oppositely directed beams we find for the energy $E_{\text{ra(cm)}}$ in the center-of-mass frame, again in the limit of large momenta (where (3.30) holds),

$$E^2_{\text{ra(cm)}} \simeq 4|\vec{p}_1|^2 c^2. \tag{8.5}$$

From (8.3) we thus find

$$E^2_{\text{ra(cm)}} \gg E^2_{\text{ft(cm)}}. \tag{8.6}$$

Hence it follows from the relativistic relation between energy and momentum that the center-of-mass energy of oppositely directed beams is very much larger than for a fixed-target experiment. Accordingly, for a given maximal momentum $|\vec{p}|$ of beam particles, the production of new (heavy) particle–antiparticle pairs is much easier in a ring accelerator.

8.2 The Layout of Ring Accelerators and Detectors

In this section we will sketch the functioning of ring accelerators and detectors. We should mention, however, that fixed-target experiments are still of relevance: in fixed-target experiments we can produce beams of unstable, but long-lived particles such as pions and muons, which can subsequently hit a "secondary target", which allows their scattering processes to be studied. Moreover, beams of stable antiparticles such as positrons and antiprotons can be produced in fixed-target experiments and be fed into ring accelerators subsequently.

Up to now experiments have been carried out at electron–positron, electron–proton, proton–proton, and proton–antiproton ring accelerators. The most energetic proton–antiproton ring accelerator was the Tevatron at Fermilab near Chicago, in which proton and antiproton beams were accelerated to an energy of 980 GeV each (i.e., a total energy of 1.96 TeV). The most energetic accelerator at present is the proton–proton ring accelerator LHC at CERN near Geneva of a (planned) total energy of 14 TeV (7 TeV per beam) in the center-of-mass frame; presently it is running at half of the foreseen total energy. The most powerful electron–positron ring accelerator was the Large Electron Positron Collider (LEP) at CERN in a tunnel of 27 km circumference 50–175 m underground and of a maximal energy, after various upgrades, of

104 GeV per electron and positron (i.e., a total energy of 208 GeV). Today the LEP tunnel is used for the LHC.

Two kinds of forces have to act on particles in ring accelerators:

(i) the particles must be accelerated, and
(ii) the particle trajectory must be bent in the form of an approximate circle inside the ring accelerator.

For both forces the Lorentz force (5.1) is employed, which means only electrically charged particles can be used. For the acceleration, an electric field $\vec{E}$ is set up essentially parallel to the beam in so-called cavities. At the LHC, eight such cavities are positioned along the approximately circular tube, which generate an electric field of about 5 MV/m (5×10^6 volts per meter!).

The beams are not continuous; the particles are accelerated in so-called bunches. A bunch is a few centimeters long, only a few 10^{-2} mm wide in the interaction points, and contains about 10^{11} protons at the LHC. The cavities serve also to compress the diameter of the bunches. The aim is to fill each beam with 2808 bunches. The electric field in the cavities oscillates with a frequency of about 400 MHz (4×10^8 times per second) such that it accelerates the bunches always in the correct direction: during each passage of a cavity, a proton receives an energy push of up to 2 MeV.

However, in ring accelerators particles lose energy by synchrotron radiation: particles on circular trajectories are subject to an acceleration pointing towards the center of the circle. Accelerated electrically charged particles always emit photons, i.e., electromagnetic waves; in the case of electrically charged particles on circular trajectories this radiation is called *synchrotron radiation*. (Synchrotrons were the first circular particle accelerators.) Hence, electrically charged particles on circular trajectories continually lose energy.[1]

For a particle of charge $\pm e$, mass m, and energy E on a circular trajectory of radius R, the energy loss per unit time interval (i.e., the radiated power P) is

$$P = \frac{ce^2}{6\pi\varepsilon_0 R^2} \left(\frac{E}{mc^2}\right)^4, \tag{8.7}$$

where ε_0 is the permittivity of the vacuum given in (5.8).

Thus, unfortunately, the energy loss increases with the fourth power of the particle energy, which limits the maximal possible particle energy for a given radius R of the circle. In order to reduce this energy loss we are forced to chose the radius R as large as possible; this explains the enormous circumference of today's ring accelerators. In addition, the loss is larger, the smaller the mass of the accelerated particles, and hence particularly large for electron–positron ring accelerators (see the exercise at the end of this chapter) such as LEP at CERN. However, the construction of even larger

---

[1] The fact that the electrons of the atomic shells do not emit synchrotron radiation shows again that the idea of a point-like electron from Sect. 5.5 cannot be correct. The description of an electron in terms of a standing wave in the framework of the Bohr atomic model as in Fig. 5.15, for which only given values of the energy are allowed, removes this problem.

Fig. 8.4 Schematic layout of
a ring accelerator

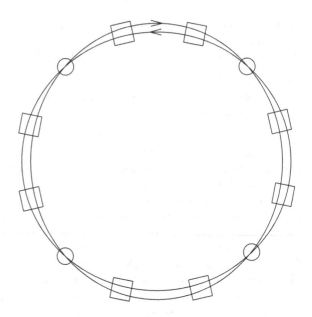

electron–positron ring accelerators is practically impossible; if we want to study
electron–positron collisions at still larger energies in the future we have to consider
linear accelerator facilities. Such a project, the International Linear Collider, ILC, of
a length of about 35 km, is at present under discussion.

The required curvature of particle trajectories in ring accelerators is achieved with
the help of vertically applied magnetic fields $\vec{B}$: inside a magnetic field the Lorentz
force (5.1) acts in a direction perpendicular to both the velocity and the magnetic
field. Of course the magnetic field is poled such that the corresponding force points
towards the inside of the horizontal circular trajectory. To this end we need, at the
LHC, about 5000 superconducting magnets, in which magnetic fields are produced
of up to $8.4\,T = 8.4\,kg/(Cs)$, where T is tesla. (For comparison: the Earth's magnetic
field has a strength of about $5 \times 10^{-5}\,T$.) For the generation of these magnetic
fields we need currents of $\sim 11700\,A$, which flow inside superconducting cables
at a temperature of $\sim 1.9\,K$ ($\sim 1.9°$ above absolute zero). To cool all $7600\,km$ of
superconducting cables we need about $700,000\,l$ of liquid helium! In September
2008 a welded joint of such a cable did not bear the charge. It exploded, destroying
a tank of helium, the explosion of which, in turn, displaced a magnet.

The particle beams cross at interaction points, which are surrounded by detectors.
At the LHC four such interaction points exist, with the four detectors ALICE, ATLAS,
CMS, and LHCb (which are specialized for different tasks); hence its construction
resembles Fig. 8.4, where we have omitted the magnets. The circles in Fig. 8.4 denote
the interaction points, and the squares the cavities.

Next we turn to how the detectors function. Here we take advantage of the
following phenomenon: when particles pass through matter, they scatter off atoms;
charged particles and photons scatter off the electrons of the atomic shells or, in

particular for strongly interacting particles, i.e., hadrons, off atomic nuclei. In this process energy and momentum are transferred to the electrons and nuclei, and the atoms are destroyed. Hence, a track of unbound electrons is produced along the trajectory of flight of particles through matter; the number of these free electrons is proportional to the energy loss of the particle and hence to its energy.

These free electrons are collected in detectors: often one uses gas-filled detectors of layers of metallic plates within which layers of parallel wires are arranged. An electric potential difference on the order of kilovolts is applied between the plates and the wires; then the free electrons move onto the wires and create a current pulse. We try to detect their original number, their position, and their time of creation as precisely as possible in order to reconstruct the track and the energy of the particle. In the case of (multiwire) proportional chambers, the applied potential difference is chosen such that the measured current pulse is proportional to the energy of the particles to be measured; the 1992 Nobel prize was awarded to G. Charpak for their development. Apart from gas-filled detectors there also exist detectors consisting of solids such as semiconductors, which can also be used to collect free electrons. In so-called scintillators (often of plastic) charged particles generate numerous photons (quanta of light), which can be detected in photocathodes.

In addition a magnetic field is applied inside the apparatus. As already discussed in the context of ring accelerators, the Lorentz force acts on a moving particle inside a magnetic field perpendicular to its direction of flight. Then a particle of charge q, with velocity directed perpendicular to the magnetic field, moves on a circular trajectory of radius R,

$$R = \frac{p}{qB}, \tag{8.8}$$

where p is the modulus of its momentum and B the modulus of the magnetic field. Thus the measurement of the curvature of the track of a particle allows its momentum to be determined. (As a result of scatterings off atoms, its momentum and hence, according to (8.8), the radius of curvature R of its trajectory decrease along its trajectory, i.e., the trajectory bends more and more.) Finally we can determine the time of flight and thus the velocity of particles with long enough lifetime. All this information serves to reconstruct the properties—momenta, energy, mass, charge, interactions—of all particles produced at the interaction point as precisely as possible.

Roughly speaking, we can distinguish four kinds of particles produced at the interaction point:

(i) Particles that decay very rapidly by the strong interaction, such as the Δ baryons in Chap. 6; only their decay products are visible in the detector.

(ii) Particles that decay slowly by the weak interaction, such as muons, τ leptons, and hadrons made out of unstable s, c, b, or t quarks; these decay often away from the interaction point (with the exception of the too rapidly decaying t quarks) but still inside the detector (with the exception of relatively long-lived muons).

(iii) Stable particles, such as electrons, positrons, protons, neutrons, and photons.

Fig. 8.5 Possible final state after an inelastic scattering process

(iv) Invisible particles that interact neither strongly nor electromagnetically; among the known particles these are the neutrinos.

Hence a detector consists of different components: near to the interaction point one finds the so-called vertex detector (or tracker), of high spatial resolution, often a semiconductor. It allows tracks of particles to be measured with a precision of a few hundredths of a millimeter, and thus to detect vertices originating from particles decaying slightly away from the interaction point.

Around the tracker lie the so-called calorimeters, whose purpose is, above all, to measure the energy: first there are the electromagnetic calorimeters, which are essentially sensitive to electrons and photons, then the hadron calorimeters, which are sensitive mostly to strongly interacting particles (hadrons). All these modules are positioned in cylindrical layers around the interaction point and the beam axis. Furthermore they are inside extremely strong magnetic fields (of 2–4 T at the LHC) in order to measure the momenta of the particles, in addition to their energies.

Practically no particle passes through all these materials; exceptions are muons, which do not interact strongly and which can traverse more material than electrons owing to their larger mass. Hence the outermost layers are the so-called muon chambers: if a particle track is found there it is—with a large probability—a muon. Nowadays the whole apparatus is the size of a house. (The ATLAS detector at the LHC is 46 m long and 25 m high.)

A scattering event can lead, e.g., to the tracks shown in Fig. 8.5, where we have omitted the incoming particle beams; we know only that they intersected within the interaction point represented by the black circle.

In Fig. 8.5 we can identify the particles with tracks that curve strongly near the end as two beams of hadrons directed to the left and to the right (such beams are denoted as *jets*) originating probably from a quark–antiquark pair. The particle with the very long track on the right is probably a muon, originating possibly from the weak decay of, e.g., a b quark (see Fig. 7.4). Here we see ten particle tracks; at the LHC one event can lead to up to 1000 particle tracks!

At the interaction points, two bunches cross about every 25 ns (2.5×10^{-8} s); each bunch crossing can lead to up to 20 collisions of single protons. We are interested mostly in the more interesting inelastic scattering events, but even these occur up to 6×10^8 times per second!

Clearly, this leads to an enormous problem of data handling: each of the up to 6×10^8 scattering events per second generates a large amount of information in the various modules of the detectors, and it is impossible to store it all. The only way out lies in an early selection of data depending on the nature of the event. For this purpose we use so-called trigger systems: triggers just check the rough features of an event and decide automatically whether it is interesting enough for all the data to be saved. To this end we have to chose criteria for the triggers that allow them to separate interesting from uninteresting events. Typical such criteria are

(i) the presence of jets or leptons with very large momenta, which could originate from the decay of very massive particles as W, Z, or Higgs bosons, but also from other new particles;
(ii) nonzero momentum balance of all momentum components perpendicular to the beam axis: owing to momentum conservation, the sum of all of these momentum components should in fact be zero. If this is not the case, invisible particles such as neutrinos or new species of invisible particles have been produced. Indeed, dark matter in our Universe consists possibly of such invisible new particles!

Although such triggers unblock often only one in 10^7 events for complete storage of all data, about 1.8 Gbyte of data accumulate per second at the LHC, which, of course, cannot be analyzed immediately. We expect that, with the help of clusters of computers, about 10^9 events can be analyzed per year; accordingly it can take months until discoveries hidden in the data are found (and verified). However, this phenomenon is well known from earlier experiments in particle physics.

8.3 The Search for New Elementary Particles

Before we turn to the search for new particles at the LHC, we will sketch how new particles are searched for in electron–positron collisions. Here the underlying process is relatively simple and represented in Fig. 8.3. We have already discussed the condition on the center-of-mass energy ($E > 2m_\mathrm{p}c^2$) that must be satisfied in order that a particle–antiparticle pair of mass m_p can be produced. If this condition is satisfied, the computation of the probability of production of a particle–antiparticle pair has to be carried out using the Feynman rules sketched in Chap. 5.

According to these rules the amplitude whose square gives the probability is the product of several factors. Each vertex connecting a photon to a charged particle contributes to these factors with one power of the coupling constant g (or the charge q of the corresponding particle, see (5.26)). It follows that the probability of production of a given particle–antiparticle pair in inelastic electron–positron scattering is proportional to

(i) the square of the electron charge e^2,
(ii) the square of the charge of the produced particle p, i.e., q_p^2,
(iii) the square of a function $f(E, m_\mathrm{e}, m_\mathrm{p})$ depending on the photon propagator (i.e., the photon energy $E = E(\mathrm{e}^+) + E(\mathrm{e}^-)$) and the masses of the particles in the initial and final states. (Here we are not interested in the angle θ under which

the particles p and $\bar{\text{p}}$ are emitted; we assume that we have integrated over all possible angles θ.)

The function $f(E, m_e, m_p)$ vanishes for $m_p > E/(2c^2)$, where it is not possible for the energy to be conserved. For $m_p \ll E/(2c^2)$, on the other hand, it is practically independent of the mass m_p of the produced particle. Then we obtain for the production probability of a particle–antiparticle pair in electron–positron scattering

$$P(\text{e}^+ + \text{e}^- \rightarrow \text{p} + \bar{\text{p}}) \simeq e^2 \times q_\text{p}^2 \times f^2(E), \tag{8.9}$$

where we have omitted the dependence of f on the electron/positron mass in the initial state.

Then we find for the production of μ or τ leptons (because of $q_\mu = q_\tau = e$, and for a total energy E much larger than the μ or τ lepton masses)

$$P(\text{e}^+ + \text{e}^- \rightarrow \mu^+ + \mu^-) \simeq P(\text{e}^+ + \text{e}^- \rightarrow \tau^+ + \tau^-) = e^2 \times e^2 \times f^2(E). \tag{8.10}$$

The result for $P(\text{e}^+ + \text{e}^- \rightarrow \text{e}^+ + \text{e}^-)$ is somewhat different since here the first of the diagrams in Fig. 8.2 contributes as well.

Hadrons can also be produced in electron–positron collisions. We know that hadrons consist of quarks. The production of hadrons proceeds in two steps: first, a quark–antiquark pair is produced as in Fig. 8.6. Subsequently the quarks and antiquarks emit (virtual) gluons, which decay into additional quark–antiquark pairs, and finally the quarks and antiquarks form hadrons such as pions, protons, or neutrons. (Free quarks do not exist owing to confinement.) This second stage is denoted as *hadronization*. If the original quarks and antiquarks were very energetic, the generated hadrons fly along the direction of the original quarks or antiquarks in the form of clustered jets.

In order to compute the probability of the production of hadrons, we can make the simplifying assumption that it depends only on the probability of the production of the first quark–antiquark pair; the probability of hadronization is always equal to one if we sum over all possibilities to produce various hadrons.

This assumption allows the use of (8.9) for the computation of the probability of the production of hadrons—under the condition that the mass of the initially produced quarks is smaller than $E/(2c^2)$, and that we sum over all quarks with this property.

For several reasons it is useful to consider the ratio of the probabilities of the production of hadrons to the production of muons. In contrast to electrons, muons are relatively easy to identify, and measurements of ratios are independent of the number of particles in the beams, the duration of the measurement, and many properties of the particle detectors. Furthermore (8.9) gives a simple expression for this ratio denoted by R:

$$R = \frac{P(\text{e}^+ + \text{e}^- \rightarrow \text{Hadrons})}{P(\text{e}^+ + \text{e}^- \rightarrow \mu^+ + \mu^-)} = \sum_i \frac{e^2 \times q_i^2 \times f^2(E)}{e^2 \times e^2 \times f^2(E)} = \sum_i \left(\frac{q_i}{e}\right)^2, \tag{8.11}$$

where we have to sum over all quarks lighter than $E/(2c^2)$. This leads to a dependence of R on the total energy E.

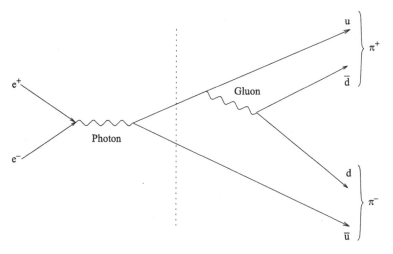

Fig. 8.6 Production of hadrons via a quark–antiquark pair by electron–positron annihilation

The ratio R has been measured in many electron–positron collision experiments in different regimes of the total energy E. In Fig. 8.7 we show the results in the domain $2\,\text{GeV} < E < 15\,\text{GeV}$—together with their error bars—of the following experiments:

$2\,\text{GeV} < E < 4.8\,\text{GeV}$: BES Collaboration [21] at the Beijing Electron Positron Collider (BEPC) in Beijing (circles)
$5\,\text{GeV} < E < 7.4\,\text{GeV}$: Crystal Ball Collaboration [22] at the SPEAR storage ring in Stanford, USA (stars)
$7.4\,\text{GeV} < E < 9.4\,\text{GeV}$: LENA Collaboration [23] at the DORIS storage ring at DESY, Hamburg (squares)
$7.3\,\text{GeV} < E < 10.4\,\text{GeV}$: MD-1 detector [24] at the VEPP-4 collision experiment in Novosibirsk, USSR (circles)
$12\,\text{GeV} < E < 15\,\text{GeV}$: five different detectors at the PETRA storage ring at DESY, Hamburg: Tasso [25, 26] (squares), Jade [27] (circles), Pluto [28] (tilted squares), Mark J [29] (triangles), and Cello [30] (star).

What would the result according to the formula (8.11) look like? Let us consider first the interval $2\,\text{GeV} \lesssim E \lesssim 4\,\text{GeV}$. Here the following quarks are lighter than $E/(2c^2)$: the u quark (with $q_u = \frac{2}{3}e$), the d quark (with $q_d = -\frac{1}{3}e$), and the s quark (with $q_s = -\frac{1}{3}e$). The c quark is not light enough for pair production. Thus the sum over the squares of the charges gives $R = (2/3)^2 + (1/3)^2 + (1/3)^2 = 6/9 = 2/3$.

However, the measured value of R is somewhat larger than 2, hence too large by about a factor of 3! This discrepancy continues for $4\,\text{GeV} \lesssim E \lesssim 10\,\text{GeV}$, where in addition the c quark (with $q_c = \frac{2}{3}e$) can be produced, and in the interval $E \gtrsim 12\,\text{GeV}$, where the b quark (with $q_b = -\frac{1}{3}e$) contributes as well.

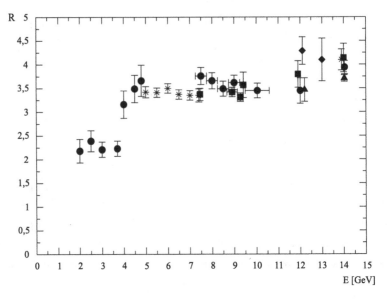

Fig. 8.7 Results of measurements of the ratio R, defined in (8.11), as a function of the energy E

However, we forgot the color in our computation of R: according to QCD each quark u, d, s, etc. carries one of three possible colors, and in the sum over i in (8.11) we have to take into account each of the three colors of a quark separately—every quark of a given color contributes separately to this sum. Therefore the prediction for R is three times that in the case of "colorless" quarks: $R \sim 2$ in the interval $2\,\text{GeV} \lesssim E \lesssim 4\,\text{GeV}$, $R \sim 3\frac{1}{3}$ in the interval $4\,\text{GeV} \lesssim E \lesssim 9\text{–}10\,\text{GeV}$, and $R \sim 3\frac{2}{3}$ for $E \gtrsim 12\,\text{GeV}$.

Now the computation agrees approximately with the results of the measurements. Two points should still be noted, however:

(a) The results of measurement are systematically somewhat larger than our prediction, where we have neglected the effect of hadronization. Indeed one can show that hadronization—at least the emission of a gluon by one of the produced quarks—should imply a slight increase of R, which corresponds to the observations. This increase of R depends in a calculable way on the strong fine structure constant α_s and can thus be used for a measurement of α_s, see Fig. 11.5.

(b) In Fig. 8.7 we have omitted results of measurements in the intervals $3\,\text{GeV} \lesssim E \lesssim 4\,\text{GeV}$ and $9.5\,\text{GeV} \lesssim E \lesssim 10.5\,\text{GeV}$, where R varies strongly: if the total energy E is close to a meson mass, R increases strongly, since the two produced quarks first form a bound state (the corresponding meson), which then decays subsequently. In the interval $3\,\text{GeV} \lesssim E \lesssim 4\,\text{GeV}$ these are the J/Ψ mesons, which consist of a c and a $\bar{c}$ quark, and for $9.5\,\text{GeV} \lesssim E \lesssim 10.5\,\text{GeV}$ these are the Υ mesons, consisting of a b and a $\bar{b}$ quark. These quarks were discovered by the production of the corresponding mesons. (The 1976 Nobel prize was

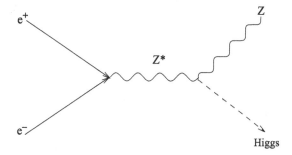

Fig. 8.8 Possible production of a Higgs boson (together with a Z boson) via a virtual Z boson in electron–positron annihilation

awarded to B. Richter and S. Ting for the independent discovery of the c quarks in J/Ψ mesons at the Stanford Linear Accelerator Center (SLAC) and at the Brookhaven National Laboratory (BNL) in 1974.)

Above all, the above-mentioned color factor three in the computation of R contributed to the proof that each quark exists in three versions—carrying one of the three possible colors—which is very difficult to verify otherwise.

The most energetic electron–positron collision experiments to date have been carried out at LEP at CERN. In fact no additional quark or an additional lepton was discovered, but the properties of the Z boson—its mass, its couplings, and its decay probabilities into the various quarks and leptons—were measured to high precision. The results of these measurements agree very well with the predictions in the framework of the theory of the weak interaction in Chap. 7.

An additional aim of the experiments at LEP was the discovery of the Higgs boson. Higgs bosons can be produced in electron–positron collisions via the Feynman diagram in Fig. 8.8. In Fig. 8.8, Z* denotes a virtual Z boson; Z and Higgs stand for real particles satisfying the energy–momentum relation (3.22).

The Feynman diagram in Fig. 8.8 is not quite complete, since both the real Z boson and the real Higgs boson are unstable: a Z boson can decay into all of the quark–antiquark or lepton–antilepton pairs (apart from top–antitop quarks, which are too heavy). Likewise a Higgs boson can decay into all of the quark–antiquark, lepton–antilepton, or boson–antiboson pairs that are light enough. However, the relative decay probabilities are always proportional to the squares of the corresponding couplings. The couplings of a Higgs boson to quarks and leptons are the Yukawa couplings (7.20), and hence proportional to the masses of the quarks and leptons. Accordingly the Higgs boson decays preferably (i.e., with the largest probability) into the heaviest pair of particles; owing to energy conservation, the mass of these particles must be smaller than half the Higgs mass, however. Correspondingly, if its mass is smaller than about $140 \, \text{GeV}/c^2$, the Higgs boson decays preferably into a b–b̄ pair. (In the case of a Higgs mass larger than about $140 \, \text{GeV}/c^2$, the Higgs boson decays preferably into a W^+–W^- pair, where, for a Higgs mass below $160 \, \text{GeV}/c^2$, one of the W bosons is virtual.)

It follows from energy conservation, similar to the inequality (8.1), that the process in Fig. 8.8 is possible only if the sum of the Higgs and the Z boson masses is smaller

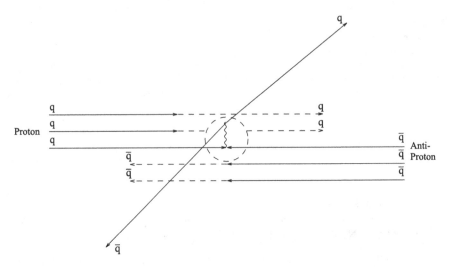

Fig. 8.9 Schematic representation of a proton–antiproton collision where, inside the dashed circle representing a magnifying glass, only one quark of a proton annihilates with an antiquark of an antiproton into a boson. Here, the boson decays into a quark–antiquark pair

than the total energy E/c^2. Thus, for $E \sim 208\,\text{GeV}$ and with $M_Z \sim 91\,\text{GeV}/c^2$, a Higgs boson could have been discovered at LEP only if it is lighter than about $114\,\text{GeV}/c^2$.

However, no Higgs boson was observed at LEP, from which we conclude that it is heavier than about $114\,\text{GeV}/c^2$. (This conclusion holds only under the assumption that the coupling of the Higgs boson to the Z boson, which appears in Fig. 8.8, and also the Higgs decay probability into b quarks correspond to the usual expectations; if these quantities are smaller, a lighter Higgs boson would not necessarily have been observed.)

Now we turn to the search for new particles in proton–proton and proton–antiproton ring accelerators. As a result of the larger mass of the protons, their energy loss by synchrotron radiation (8.8) is much smaller. Thus they can be accelerated to much higher energies than electrons and positrons.

The disadvantage of proton–proton or proton–antiproton collisions is, however, that these processes are much more complicated than those represented in Fig. 8.3, since protons consist of three quarks (and antiprotons of three antiquarks). In an energetic proton–antiproton collision, only one quark of a proton can annihilate with an antiquark of an antiproton into a boson (gluon, photon, W or Z boson), which decays subsequently into a particle–antiparticle pair; see Fig. 8.9, where we have omitted the process of hadronization. The other quarks practically do not contribute to the energy of the subprocess inside the dashed circle.

A quark inside an energetic proton does not necessarily carry a third of its kinetic energy: it is indeed true that the sum of the energies of the three quarks must be equal to the proton energy, but the energy of a single quark can vary between 0 and the total

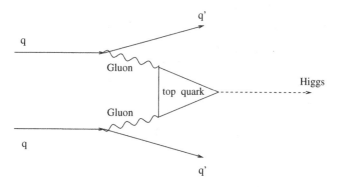

Fig. 8.10 Production of a Higgs boson from two quarks q via gluon fusion

energy of the proton. Thus, in contrast to the energies of the electrons and positrons in Fig. 8.3, the energies of the quarks and antiquarks participating in the subprocess inside the circle in Fig. 8.9 are not known. This—and the larger number of hadrons in the final state from the "left over" quarks and antiquarks—makes the study and the interpretation of proton or antiproton collisions much more complicated.

The most energetic proton antiproton accelerator was the Tevatron at Fermilab near Chicago. The top quark with its mass of about $173 \, \text{GeV}/c^2$ was discovered at this accelerator, and the existence of the W boson was confirmed. The Higgs boson was also searched for, but the sensitivity of the Tevatron depends quite strongly on the unknown Higgs boson mass. According to the latest results, a Higgs boson mass in the range of about 155–$175 \, \text{GeV}/c^2$ is very unlikely.

In the following years our attention will turn to the LHC, with its much larger center-of-mass energy. This energy should suffice to reveal the Higgs boson in (nearly) every case—unless it is unusually heavy (heavier than about $1 \, \text{TeV}/c^2$) or if it decays very unusually, which would hinder its detection.

However, the production processes for a Higgs boson in proton–proton collisions are more complicated than in electron–positron collisions such as in Fig. 8.8. Similar to the proton–antiproton collisions in Fig. 8.9, only two quarks q (one from each proton) contribute to the subprocess. In Figs. 8.10 and 8.11 we have sketched two dominant processes at the "quark level", relevant for the production of a Higgs boson at the LHC. (In contrast to Fig. 8.9, where the horizontal axis corresponds to the beam axis, in Figs. 8.10 and 8.11 the time axis runs from left to right.)

The direct couplings of u and d quarks to the Higgs boson are negligibly small, since these couplings are proportional to their very small quark masses. In Fig. 8.10, each of the quarks q (u or d quark) first emits a gluon. Still, gluons do not couple directly to Higgs bosons (which carry no color), but only to heavy quarks such as the t quark. The coupling of t quarks to Higgs bosons is particularly strong, since t quarks are particularly heavy. Hence two gluons can produce a Higgs boson only indirectly via a so-called "loop" (here: a triangle) of virtual t and anti-t quarks. In principle, contributions from other quarks in the loop exist as well, but these are

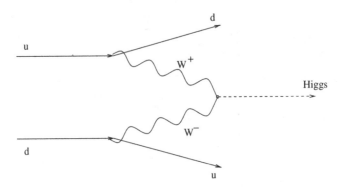

Fig. 8.11 Production of a Higgs boson from two quarks u and d via vector boson fusion, here the fusion of W bosons

relatively small owing to their smaller couplings to the Higgs boson. The quarks q′ are the same as the quarks q, but with different colors after the emission of gluons. The complete process in Fig. 8.10 is denoted as Higgs production via "gluon fusion".

In Fig. 8.11, each of the u and d quarks emits first a W boson. In contrast to gluons, W bosons couple directly to the Higgs boson. In principle there exists a similar process where the W bosons are replaced by Z bosons (and the nature of the incoming quarks remains unchanged). Since W and Z bosons are so-called vector bosons with spin $\hbar$, the process in Fig. 8.11 is denoted as Higgs production via "vector boson fusion".

However, it is not sufficient to produce Higgs bosons; they must be detected as well. Already in our discussion of Fig. 8.8 we emphasized that Higgs bosons are not stable and decay, depending on their mass, mainly into a b–$\bar{\text{b}}$ pair (if lighter than about $140\,\text{GeV}/c^2$) or into a W+–W− pair (if heavier than about $140\,\text{GeV}/c^2$). In addition, neither the b quarks nor the W bosons are stable, which leads to complicated final states after the decay of a Higgs boson.

How can we tell that such a final state originates from the decay of a Higgs boson? To this end it is useful to consider first a two-particle final state: let us assume that a Higgs boson of mass M_H decays into two particles, and that we can measure their energies E_1, E_2 and momenta $\vec{P}_1$, $\vec{P}_2$. As a result of energy and momentum conservation, this allows us to deduce the energy E_H and the momentum $\vec{P}_\text{H}$ of the Higgs boson:

$$E_\text{H} = E_1 + E_2, \quad \vec{P}_\text{H} = \vec{P}_1 + \vec{P}_2. \tag{8.12}$$

Now E_H and $\vec{P}_\text{H}$ satisfy (3.24), i.e.,

$$\begin{aligned} M_\text{H}^2 c^4 &= E_\text{H}^2 - \vec{P}_\text{H}^2 c^2 \\ &= (E_1 + E_2)^2 - (\vec{P}_1 + \vec{P}_2)^2 c^2. \end{aligned} \tag{8.13}$$

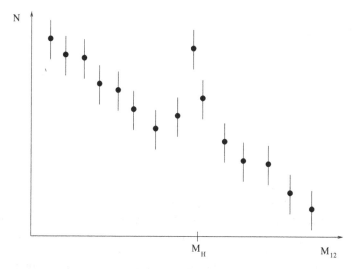

Fig. 8.12 Possible result for number of events N as a function of invariant mass M_{12}

Thus the energies and momenta of the two particles always satisfy the relation (8.13), if they originate from the decay of a Higgs boson. However, two such particles can also originate from completely different processes that have nothing to do with the decay of a Higgs boson. In this case their energies and momenta will generally *not* satisfy the relation (8.13).

This reasoning can easily be generalized to more complicated final states with more particles, if all their energies and momenta can be measured. Since we do not know the Higgs mass beforehand, we can proceed as follows (assuming a two-particle final state).

(a) We concentrate on two particle species that could originate, in principle, from the decay of a Higgs boson. If they appear in an event, we measure their energies and momenta, from which we can compute the quantity $M_{12}^2 = (E_1+E_2)^2 c^{-4} - (\vec{P}_1 + \vec{P}_2)^2 c^{-2}$. This quantity is also called the "invariant mass", since it does not depend on the reference frame used.

(b) We plot the number N of measured values of M_{12} as a function of M_{12}.

Events that have nothing to do with the decay of a Higgs boson, the so-called background, will lead to a regular (often decreasing) distribution of N with M_{12}. However, events that originate from the decay of a Higgs boson will always give the same value for M_{12} ($\pm$ measurement errors). Then N as a function of M_{12} could look as sketched in Fig. 8.12, for example.

Now we look for values of M_{12} that, compared to the background, are particularly frequent. In Fig. 8.12 we see such a peak, which we explain by the occasional production of Higgs bosons of corresponding mass (denoted by M_{H} in Fig. 8.12). It is not always easy to decide whether such peaks are hints of or even proofs of

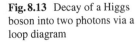

Fig. 8.13 Decay of a Higgs boson into two photons via a loop diagram

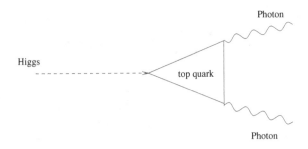

the existence of new particles; to this end, precise analyses of possible measurement errors and statistical fluctuations are necessary.

In the concrete case of the search for Higgs bosons at the LHC, the situation is complicated by the fact that the invariant masses of the decay products $b-\bar{b}$ or W^+-W^- are difficult to determine. In addition, numerous $b-\bar{b}$ pairs are produced by processes involving the strong interaction. This makes it particularly difficult to detect peaks in the distribution of invariant masses of $b-\bar{b}$ pairs corresponding to a Higgs boson.

One way out consists in concentrating on rare Higgs decays, whose final states are particularly easy to analyze, however. An example is the decay of Higgs bosons into two photons. At first such a decay seems impossible, since a Higgs boson does not couple to photons; however, as in the case of the production of Higgs bosons via gluons as in Fig. 8.10, such a coupling can be generated by a loop diagram as in Fig. 8.13.

The probability of such a decay is very small indeed—only about one in a thousand Higgs bosons will decay into two photons this way—but the energies and momenta of photons can be measured quite precisely. In particular in the case of a relatively light Higgs boson (not much heavier than the lower LEP bound of about $114\,\text{GeV}/c^2$), the analysis of the invariant masses of two photons is a promising method to detect a Higgs boson and to measure its mass. However, preferably all decay channels of a Higgs boson (e.g., into Z boson pairs and τ lepton pairs, in addition to the ones mentioned above) should be analyzed together. This way, hints for a Higgs boson with a mass below about $145\,\text{GeV}/c^2$ have been observed in data available in August 2011.

Finally we have to emphasize that, besides the Higgs boson, we hope for the discovery of additional new particles at the LHC, such as the ones predicted within the still speculative supersymmetric extension of the Standard Model, see Sect. 12.2.

Apart from these experiments at the most energetic accelerators, experiments are being carried out at accelerators of lower energy but particularly high particle densities. These serve to measure rare decays of known particles such as b quarks (in so-called "B factories"). Rare decays can be induced by virtual new particles, too heavy to be directly visible in today's accelerators.

In addition there exist many experiments in particle physics that do not depend on accelerators. We have pointed out already, near the end of Sect. 7.5, that some experiments study the properties of neutrinos, in particular the transformation of one neutrino species into another. The origin of these neutrinos can be astrophysical processes (e.g., supernovae), leading to cosmic neutrinos, nuclear reactions in the Sun (solar neutrinos), muon decays in the atmosphere (where the muons originate from scattering processes of cosmic radiation), leading to atmospheric neutrinos, nuclear reactors, or fixed-target experiments performed for this purpose.

Since interactions of neutrinos in detectors are very unlikely (since they can be induced only by the weak interaction), neutrino detectors must be extremely well protected from natural radiation in order to be sensitive to the rare events induced by neutrinos. Hence they are located in mines or tunnels, preferably several kilometers below the surface of the Earth. Another possibility is the use of seawater in the oceans (at about 1 km depth) as a detector, where interactions of neutrinos can induce weak flashes, which are detected by photon detectors hanging on long ropes.

In recent years, experiments in astroparticle physics, which specializes in the study of cosmic radiation, have become more and more important. Cosmic radiation consists of photons, electrons, positrons, protons, and antiprotons, which can be extremely energetic (up to $10^{20}\,\mathrm{eV} = 10^{11}\,\mathrm{GeV}$); such energies are unattainable in accelerators. For their study we can send detectors by balloons into the stratosphere, place them on satellites or in the International Space Station. In addition, particularly energetic (but also particularly rare) cosmic particles can generate weak lightnings in the upper atmosphere, which can be searched for by telescopes or kilometer-wide arrays of telescopes such as the Pierre Auger observatories in deserts in Argentine and Colorado.

Cosmic radiation can originate from supernova explosions, pulsars, black holes, or other astrophysical phenomena. Another possible origin is related to dark matter: if dark matter consists of particles, these particles can collide inelastically and produce other particles, which contribute to cosmic radiation. Such processes would be particularly frequent near the centers of galaxies, where the dark matter density is particularly large, and lead to particles of a well-defined energy. Once other astrophysical phenomena can be excluded for such components of the cosmic radiation (still to be discovered), we can talk about an indirect detection of dark matter.

Finally, attempts are being made to detect dark matter directly on Earth. In fact the corresponding particles would be present everywhere, but would be neutral, like neutrinos, and interact only very rarely. These particles could scatter off atomic nuclei and transfer some energy (about 10 keV); we expect, however, just about one scattering event per kilogram of material every ten days! Researchers are trying to measure this energy transfer in extremely sensitive detectors consisting of large amounts (up to 100 kg) of heavy nuclei such as germanium or xenon. As in the case of neutrino experiments, these detectors must be well protected from natural radiation and are located in the same remote places. Some neutrino experiments can be used simultaneously for the search for dark matter.

Since the particles that make up dark matter could also be produced at accelerators, we can hope for their multiple confirmation in the corresponding detectors, indirectly in cosmic radiation, and in particle physics experiments. In any case the "interaction" between traditional particle physics and the relatively new field of astroparticle physics will play an important role in the future.

Exercise

8.1. Compute the power P emitted by synchrotron radiation [see (8.8)]

(a) of an electron at LEP ($E \simeq 104 \, \text{GeV}$, circumference about 27 km) and

(b) of a proton at the LHC ($E \simeq 7 \, \text{TeV}$, same circumference).

Give the result in watts (= J/s) and in eV/s. Compare the energy loss per second of an electron (in (a)) and a proton (in (b)) to their total energy E.

Chapter 9
Symmetries

External and internal symmetries are concepts essential for the formulation of fundamental interactions in the framework of (quantum) field theory. External symmetries correspond to the invariance of measured variables under transformations in space and time. Internal symmetries correspond to the invariance of measured variables under transformations of fields, for example, multiplication of the fields by a complex phase. The postulation of internal symmetries, which correspond to complex phases depending arbitrarily on space and time, requires the introduction of so-called gauge fields. These are associated with the particles whose exchange generates the fundamental interactions. All fundamental interactions can thus be traced back to such internal symmetries. The Higgs field plays a special role, since its constant non-vanishing value in the Universe implies a so-called spontaneous breaking of the internal symmetry related to the weak interaction.

9.1 External Symmetries

We talk about a *symmetry* in physics if, after a transformation (see below), all observable or measurable quantities remain unchanged.

The most common of such transformations are transformations in space and time (so-called external transformations): at least in empty space, the laws of physics (i.e., the equations and hence the results of measurements) remain unchanged

(a) after a translation in space,
(b) after a translation in time,
(c) after a rotation around one of the three possible axes, and
(d) after a transformation into a reference frame that is moving at constant velocity in one of the three possible directions.

A transformation of kind (a) has both a formal consequence for the equations of physics and a concrete consequence for the results of measurement processes.

U. Ellwanger, *From the Universe to the Elementary Particles*,
Undergraduate Lecture Notes in Physics, DOI: 10.1007/978-3-642-24375-2_9,
© Springer-Verlag Berlin Heidelberg 2012

The formal consequence for the equations of physics consists in the fact that the position vector $\vec{r}$ can be replaced by $\vec{r}' = \vec{r} + \vec{a}$ everywhere in the equations, where $\vec{a}$ is an arbitrary constant vector. Such a replacement implies merely a new choice of the origin of the reference frame. Hence fields $\phi(\vec{r}, t)$ are to be replaced by $\phi(\vec{r}', t)$, and derivatives with respect to the components x, y, z of $\vec{r}$ by derivatives with respect to the components x', y', z' of $\vec{r}'$. Here it is important that these derivatives remain the same. Since the fundamental equations, such as the Klein–Gordon equation (4.1) and the Eqs. (5.4), (5.5) of electrodynamics, contain only derivatives with respect to the components of $\vec{r}$ (i.e., $\vec{r}$ never appears explicitly) these fundamental equations do not change up to the arguments of the fields. (Of course it is often convenient to chose the center of a body as the origin of the reference frame such that the gravitational or electric fields generated by this body depend only on the distance $|\vec{r}|$ to the origin. Even so, the origin of the reference frame can be shifted by $\vec{a}$, whereupon the gravitational or electric fields and forces depend on $|\vec{r}'| = |\vec{r} + \vec{a}|$.)

The concrete significance of such a transformation for results of measurements depends on the fact that a translation of the complete experimental apparatus by a constant vector $\vec{a}$ gives the same results. This follows directly from the symmetry of the fundamental equations. (In the case of measurements in a field, such as the gravitational field, we have to take care, of course, that the field at the new position is the same; otherwise the field would have to be shifted as well.)

Now we can easily verify that the three additional transformations (b), (c), and (d) are also symmetries of the fundamental equations in the above sense, and allow corresponding transformations of the experimental apparatus without impact on the results of measurements.

The three transformations listed under (c) (rotations) have an interesting property: after the execution of two different rotations the final result depends, in general, on the order of the rotations. A rotation of a vector around the x-axis (by, e.g., 90°), followed by a rotation around the y-axis (by, e.g., also 90°), results in a different vector than the same two rotations in the opposite order. The results differ by a rotation around the z-axis. If different transformations are connected in this way, we talk about a (non-abelian) *group* (or symmetry group, if the transformations correspond to symmetries). In the present case the corresponding group is denoted as SO(3) (also denoted as the rotation group), i.e., the group of transformations that leave invariant the modulus of a three-dimensional vector ("O" in SO(3) stands for "orthogonal" and "S" for "special"—we obtain the non-special group O(3) if we add the reflection at the origin to the three rotations.)

In special relativity theory, the three transformations mentioned under (d) are symmetries as well, but here the final result depends, as in the case of rotations, on the order of the transformations. In addition, the final result of two transformations—one of kind (c), one of kind (d)—depends on the order as well. This implies a common symmetry group of the $3+3=6$ transformations (c) and (d). If four-dimensional space-time were an "ordinary" (Euclidean) space, the corresponding symmetry group would be SO(4), i.e., the four-dimensional rotation group. Because of the relative sign in (3.19) for the scalar product of two vectors in Minkowski space, the symmetry group is denoted as SO(3,1) instead of SO(4).

A symmetry can also be broken. On the surface of the Earth, for instance, the gravitational force acts vertically. Correspondingly the movements of an object on an inclined plane—i.e., a plane inclined around the horizontal x- or y-axis—differ from the movements on a horizontal plane: rotations around the x- or y-axes are no longer symmetries; only rotations around the z-axis still correspond to a symmetry. As the origin of this symmetry breakdown we can identify the gravitational field, i.e., the 00 component of the metric (3.34) in Chap. 3, which depends on the distance r from the center of the Earth. On the surface of the Earth, the distance to the center of the Earth corresponds to the height or the z component.

If the fundamental equations are actually symmetric, and the symmetry breaking results "merely" from the presence of a field (with minimal potential energy), we talk about *spontaneous symmetry breaking*. Somewhat later we will encounter this idea again in the context of the Higgs field.

9.2 Internal Symmetries

So-called internal symmetries play an important role in field theory. Most internal symmetries are related to the fact that fields can be complex valued, i.e., correspond to complex numbers. (This concerns essentially only matter fields corresponding to quarks, leptons, and the Higgs boson, not the electromagnetic fields in Chap. 5.)

A complex number z can be written as

$$z = x + iy, \tag{9.1}$$

where i is the "imaginary" unit, defined as the square root of -1:

$$i = \sqrt{-1}. \tag{9.2}$$

The components x and y in (9.1) (both ordinary numbers) are known as the real component and imaginary component of z. Whereas an ordinary "real" number can be represented on a one-dimensional number line, a two-dimensional plane with an x- and a y-axis is required in order to represent a complex number z.

Alternatively to (9.1), a complex number z can also be written as

$$z = |z|e^{i\theta}, \tag{9.3}$$

where $|z|$ and θ are called the modulus and phase of z, respectively. The exponential function $e^{i\theta}$ of an imaginary number $i\theta$ can be identified with $\cos\theta + i\sin\theta$. $|z|$ and θ are related to x and y in (9.1) by

$$|z| = \sqrt{x^2 + y^2}, \quad \tan\theta = \frac{y}{x}, \quad x = |z|\cos\theta, \quad y = |z|\sin\theta. \tag{9.4}$$

The complex conjugate $\bar{z}$ of z is obtained by the replacement $i \to -i$, i.e.,

$$\bar{z} = x - iy = |z|e^{-i\theta}. \tag{9.5}$$

From $e^{i\theta}e^{-i\theta} = 1$ it follows easily that

$$|z| = \sqrt{z\bar{z}}. \tag{9.6}$$

For simplicity we will consider in the following scalar fields, corresponding to particles with spin 0. Quarks and leptons are particles with spin $\hbar/2$, which have to be described by so-called spinor fields; spinor fields have several components, corresponding to the different directions of their spin. The subsequent considerations are independent of this complication, however.

A field can be a real number depending on $\vec{r}$ and t, or a complex number depending on $\vec{r}$ and t. In the latter case the field has two (independent) components: like a complex number z in (9.1), a complex field $\Phi(\vec{r}, t)$ can be decomposed into its real and imaginary parts:

$$\Phi(\vec{r}, t) = \phi_r(\vec{r}, t) + i\phi_i(\vec{r}, t), \tag{9.7}$$

or into its modulus and its phase, as in (9.3):

$$\Phi(\vec{r}, t) = |\Phi(\vec{r}, t)| e^{i\varphi(\vec{r}, t)}. \tag{9.8}$$

First, the two components $\phi_r(\vec{r}, t)$ and $\phi_i(\vec{r}, t)$ of Φ can be identified as two different particle species.

Now, at last, we can discuss the first example of a so-called "internal" symmetry. In the case of a complex field we talk about an internal symmetry if all observable or measurable quantities are left unchanged under a transformation

$$\Phi(\vec{r}, t) \rightarrow \Phi'(\vec{r}, t) = \Phi(\vec{r}, t)e^{i\Lambda}. \tag{9.9}$$

If we decompose the fields $\Phi(\vec{r}, t)$ and $\Phi'(\vec{r}, t)$ into their real and imaginary components, we see that the transformation (9.9) mixes real and imaginary components:

$$\begin{aligned}
\phi_r'(\vec{r}, t) &= \cos\Lambda\ \phi_r(\vec{r}, t) - \sin\Lambda\ \phi_i(\vec{r}, t), \\
\phi_i'(\vec{r}, t) &= \sin\Lambda\ \phi_r(\vec{r}, t) + \cos\Lambda\ \phi_i(\vec{r}, t).
\end{aligned} \tag{9.10}$$

How can we interpret such a transformation in physical terms, if $\phi_r(\vec{r}, t)$ and $\phi_i(\vec{r}, t)$ correspond to two different particle species? In the case of classical particles we can easily imagine that, in a given physical process, we replace one particle by another and verify whether measured quantities, such as forces and angular distributions in scattering processes, remain unchanged. The transformations (9.10) correspond to a partial replacement of one field by another (unless the angle Λ happens to be a multiple of $\pi/2$). We cannot visualize a partial replacement of a particle by another, however.

Here we have to recall the meaning of a field in quantum field theory: the square of a field $\phi(\vec{r}, t)^2$ is proportional to the probability of finding a particle of the corresponding species at the position $\vec{r}$ at the time t. This probability can indeed change a "little"; the probability of finding a particle of type 1 can decrease somewhat, for instance, and the probability of finding a particle of type 2 can increase somewhat. This is the situation described by the transformation (9.10), which can be compared to measured quantities.

However, a fundamental rule in quantum field theory requires that the sum of the probabilities of finding any kind of particle at the position $\vec{r}$ at time t must not be changed by internal transformations. (This fundamental rule is denoted as "conservation of probability" or "unitarity".) In fact we can easily verify that the transformations (9.9) or (9.10) satisfy this rule, since

$$|\Phi(\vec{r}, t)|^2 = \phi_{\mathrm{r}}(\vec{r}, t)^2 + \phi_{\mathrm{i}}(\vec{r}, t)^2 = \phi_{\mathrm{r}}'(\vec{r}, t)^2 + \phi_{\mathrm{i}}'(\vec{r}, t)^2 = |\Phi'(\vec{r}, t)|^2 \quad (9.11)$$

holds.

Now we want to study the behavior of the fundamental equations under such transformations. The fundamental equations are the (generally massive) Klein–Gordon equations (4.11), which have to be satisfied by the components $\phi_{\mathrm{r}}(\vec{r}, t)$ and $\phi_{\mathrm{i}}(\vec{r}, t)$ above. (However, up to now we have neglected so-called coupling terms in these equations.) Now, only if the two mass terms m^2 in the two equations for $\phi_{\mathrm{r}}(\vec{r}, t)$ and $\phi_{\mathrm{i}}(\vec{r}, t)$ are the same can the two equations be combined into a single equation for the complex field $\Phi(\vec{r}, t)$:

$$\left(\frac{\partial^2}{\partial t^2} - c^2 \left(\frac{\partial^2}{\partial x^2} + \frac{\partial^2}{\partial y^2} + \frac{\partial^2}{\partial z^2}\right) + \frac{m^2 c^4}{\hbar^2}\right) \Phi(\vec{r}, t) = 0 \quad (9.12)$$

This equation is invariant under the "internal" transformations (9.9) in the following sense: if a complex field $\Phi(\vec{r}, t)$ satisfies (9.12), the equation is also satisfied by a field $\Phi'(\vec{r}, t)$ that is obtained as in (9.9). This can easily be verified by multiplying each term in (9.12) by $e^{i\Lambda}$.

However, the complete theory is invariant under the internal transformations (9.9) only if the couplings of the components $\phi_{\mathrm{r}}(\vec{r}, t)$ and $\phi_{\mathrm{i}}(\vec{r}, t)$ to other fields (and hence the constants associated with the vertices in the Feynman rules) coincide as well. If this is the case (which can be verified experimentally), we talk about an internal symmetry of the theory.

An internal transformation as in (9.9) is denoted as a U(1) transformation ("U" for "unitary"), and in the case of a symmetry we talk about a U(1) symmetry.

An interesting generalization of such an internal symmetry plays a role for the strong interaction: we mentioned in Chap. 6 that quarks carry color, and that the color of a quark can be represented by a three-component vector $\vec{q}^{\mathrm{s}}$ in "color space". (The components q_i^{s}, $i = 1, 2, 3$, correspond to the three possible colors.) Hence, for each of the six quarks of the Standard Model, we have to introduce three fields $\Phi_i(\vec{r}, t)$, which are again complex quantities (and must be spinor fields, which is irrelevant in the following). The fact that all physical properties of these three fields are the

same corresponds again to an internal symmetry: the generalization of the internal
transformation (9.9) is now of the form

$$\Phi_i(\vec{r}, t) \rightarrow \Phi_i'(\vec{r}, t) = \sum_{j=1}^{3} U_{ij}\, \Phi_j(\vec{r}, t), \qquad (9.13)$$

where U_{ij} is a unitary 3×3 matrix with complex components. (Here unitary means
that the matrix satisfies $\sum_{j=1}^{3} U_{ij} U_{kj}^* = \delta_{ik}$, where $\delta_{ik} = 1$ for $i = k$ and $\delta_{ik} = 0$ for
$i \neq k$. U^* denotes the above-mentioned complex conjugation of each component of
U. The requirement of unitarity of the matrix U follows again from the conservation
of probability.) A transformation of the kind (9.13) corresponds to a rotation of
the vector $\Phi_i(\vec{r}, t)$ in color space such that $\sum_{i=1}^{3} |\Phi_i(\vec{r}, t)|^2$ remains unchanged. In
addition, the real and imaginary parts of the various color components of $\Phi_i(\vec{r}, t)$
get mixed.

Such a transformation and the corresponding symmetry are denoted as a U(3)
transformation and U(3) symmetry, respectively. If we require in addition that the
determinant of U is equal to 1, we talk about SU(3) ("S" for "special").

The theory of strong interactions is indeed invariant under such a SU(3) symmetry
under the condition, however, that all six three-component quark fields are transformed by the same transformation U—this follows from the couplings of the quarks
to gluons, whose fields have to be transformed as well.

In Chap. 7 on the weak interaction we introduced the weak isospin, which replaces
the "color". The role of color triplets $\Phi_i(\vec{r}, t)$, $i = 1, 2, 3$, is now played by isospin
doublets $\Phi_i(\vec{r}, t)$, $i = 1, 2$. Here the physical properties of the two components are
no longer the same, and they carry different names: in (7.1) we introduced the three
quark doublets, and in (7.3) the three lepton doublets.

In spite of the different physical properties of the two components of the isospin
doublets, the fundamental equations are in fact invariant under a SU(2) symmetry,
i.e., under transformations similar to those in (9.13), where the matrices U have to
be replaced by 2×2 matrices.

The reason for the different physical properties of the two components of the
isospin doublets can be found in *spontaneous symmetry breaking* (see above) of the
SU(2) symmetry, which follows from the presence of a constant field, which is *not*
invariant under SU(2) transformations: this is the Higgs field introduced in Sect. 7.3.
In principle there exist several Higgs fields, which also form an isospin doublet; H
in Sect. 7.3 is one of its components. If one of the components of an isospin doublet
assumes a constant non-vanishing value H, this configuration is no longer invariant
under SU(2) transformations (or rotations in isospin space). The various components
of the isospin doublets of the quarks and leptons feel this symmetry breaking via
their Yukawa couplings to the Higgs field, which manifests itself in their different
physical properties, e.g., masses.

9.3 Gauge Symmetries and Gauge Fields

Another possible generalization of the transformation (9.9) arises from the possibility
that the coefficient Λ in the exponent of the exponential function can be an arbitrary
function of $\vec{r}$ and t. Such transformations are denoted as *gauge transformations*, and
are of the form

$$\Phi(\vec{r}, t) \to \Phi'(\vec{r}, t) = \Phi(\vec{r}, t) e^{ig\Lambda(\vec{r},t)}. \tag{9.14}$$

Here g is a constant that depends on the particle species described by Φ. To begin
with, the Klein–Gordon equation (9.12) is *not* invariant under the transformations
(9.14) of the field Φ (in the sense described below (9.12)), since the derivatives with
respect to t and the various components of $\vec{r}$ act also on $\Lambda(\vec{r}, t)$ (according to the
chain rule), leading to additional terms.

In order to study a generalized Klein–Gordon equation that is invariant under
(9.14), it is helpful first to rewrite the Klein–Gordon equation (9.12) in a more elegant
way. To this end we introduce coordinates x^μ ($\mu = 0, \ldots, 3$) of four-dimensional
(Minkowski) space, where $x^{1,2,3}$ correspond to the three components x, y, z, and x^0
is related to the time t by $x^0 = ct$. (Then we have $\frac{\partial^2}{\partial t^2} = c^2 \frac{\partial^2}{(\partial x^0)^2}$.)

After division by c^2, the Klein–Gordon equation (9.12) can be written in the form

$$\left(\sum_{\mu=0}^{3} g^{\mu\mu} \frac{\partial}{\partial x^\mu} \frac{\partial}{\partial x^\mu} + \frac{m^2 c^2}{\hbar^2} \right) \Phi(\vec{r}, t) = 0, \tag{9.15}$$

where $g^{\mu\mu}$ are the components of the Minkowski metric introduced in (3.33) with
coefficients 1 and -1, describing the relative signs between the time-like and spatial
derivatives. (Strictly speaking, $g^{\mu\nu}$ is the inverse of the metric $g_{\mu\nu}$ defined in (3.33),
which, however, is the same here.)

Now the action of a derivative $\frac{\partial}{\partial x^\mu}$ on a field *after* a transformation as in (9.14)
gives an additional term:

$$\frac{\partial}{\partial x^\mu} \Phi'(\vec{r}, t) = e^{ig\Lambda(\vec{r},t)} \frac{\partial}{\partial x^\mu} \Phi(\vec{r}, t) + e^{ig\Lambda(\vec{r},t)} \Phi(\vec{r}, t) ig \frac{\partial}{\partial x^\mu} \Lambda(\vec{r}, t) \tag{9.16}$$

Owing to the last term in (9.16) (which would vanish if Λ were constant), the
derivative of $\Phi'(\vec{r}, t)$ is not simply equal to the derivative of $\Phi(\vec{r}, t)$ times $e^{ig\Lambda(\vec{r},t)}$.
If this were the case, the complete Klein–Gordon equation for $\Phi'(\vec{r}, t)$ would just
be equal to the Klein–Gordon equation for $\Phi(\vec{r}, t)$ multiplied by $e^{ig\Lambda(\vec{r},t)}$, and hence
satisfied once the equation is satisfied by $\Phi(\vec{r}, t)$.

A modification of the Klein–Gordon equation leading to invariance under transformations of the kind (9.14) requires the following: to all derivatives with respect to
x^μ (of an arbitrary quantity) must be added a product of the same quantity with the
corresponding component of the four-component vector field $A_\mu(\vec{r}, t)$ and a factor
$-ig$,

$$\frac{\partial}{\partial x^\mu} \to \frac{\partial}{\partial x^\mu} - ig A_\mu(\vec{r}, t). \tag{9.17}$$

This, by itself, is not yet sufficient; in addition the following prescription has to be employed: whenever a field $\Phi(\vec{r}, t)$ is transformed as in (9.14), $A_\mu(\vec{r}, t)$ must also be transformed according to the rule

$$A_\mu(\vec{r}, t) \to A'_\mu(\vec{r}, t) = A_\mu(\vec{r}, t) + \frac{\partial}{\partial x^\mu} \Lambda(\vec{r}, t), \tag{9.18}$$

where $\Lambda(\vec{r}, t)$ is the same function as in (9.14). $A_\mu(\vec{r}, t)$ is known as the *gauge field*.

Here we see for the first time a four-dimensional vector field $A_\mu(\vec{r}, t)$, where the index μ can assume four different values $\mu = 0, \ldots, 3$. In fact, we can identify the different components of $A_\mu(\vec{r}, t)$ with the electromagnetic fields $\phi(\vec{r}, t)$ and $\vec{A}(\vec{r}, t)$ already introduced in Sect. 5.1 (see (5.4) and (5.5)) according to the rule $A_0 = \phi/c$ and the correspondence of the "spatial" components $A_{1,2,3}$. Below we will discuss this correspondence in more detail.

First we have to clarify how the rules (9.17) and (9.18) lead to an invariance of the Klein–Gordon equation. Therefore we study first the behavior of the expression

$$\left(\frac{\partial}{\partial x^\mu} - ig A_\mu(\vec{r}, t)\right) \Phi(\vec{r}, t) \tag{9.19}$$

under the transformations (9.14) and (9.18). The first term $\frac{\partial}{\partial x^\mu} \Phi(\vec{r}, t)$ has already been investigated in (9.16), and in the second term we have to use (9.14) as well as (9.18). Altogether we obtain

$$\begin{aligned}
\left[\left(\frac{\partial}{\partial x^\mu} - ig A_\mu(\vec{r}, t)\right) \Phi(\vec{r}, t)\right]' &= \left(\frac{\partial}{\partial x^\mu} - ig A'_\mu(\vec{r}, t)\right) \Phi'(\vec{r}, t) \\
&= e^{ig\Lambda(\vec{r},t)} \frac{\partial}{\partial x^\mu} \Phi(\vec{r}, t) + e^{ig\Lambda(\vec{r},t)} \Phi(\vec{r}, t) ig \frac{\partial}{\partial x^\mu} \Lambda(\vec{r}, t) \\
&\quad - ig \left(A_\mu(\vec{r}, t) + \frac{\partial}{\partial x^\mu} \Lambda(\vec{r}, t)\right) e^{ig\Lambda(\vec{r},t)} \Phi(\vec{r}, t) \\
&= e^{ig\Lambda(\vec{r},t)} \left(\frac{\partial}{\partial x^\mu} - ig A_\mu(\vec{r}, t)\right) \Phi(\vec{r}, t), \tag{9.20}
\end{aligned}$$

since the derivatives of $\Lambda(\vec{r}, t)$ cancel. Owing to the property (9.20), the combination (9.17) of a derivative and a vector field is also denoted as a *covariant derivative* D_μ:

$$D_\mu = \frac{\partial}{\partial x^\mu} - ig A_\mu(\vec{r}, t) \tag{9.21}$$

satisfies

$$\left[D_\mu \Phi(\vec{r}, t)\right]' = e^{ig\Lambda(\vec{r},t)} D_\mu \Phi(\vec{r}, t). \tag{9.22}$$

Now it becomes apparent how the Klein–Gordon equation (9.15) has to be modified in a reasonable way: every derivative $\frac{\partial}{\partial x^{\prime\mu}}$ has to be replaced by a covariant derivative D_μ, whereupon the equation assumes the form

$$\left(\sum_{\mu=0}^{3} g^{\mu\mu} D_\mu D_\mu + \frac{m^2 c^2}{\hbar^2}\right) \Phi(\vec{r}, t) = 0. \tag{9.23}$$

After a gauge transformation—where Φ and $A_\mu(\vec{r}, t)$ in D_μ have to be transformed according to (9.14) and (9.18), respectively—(9.23) becomes with the help of (9.22)

$$e^{ig\Lambda(\vec{r},t)} \left(\sum_{\mu=0}^{3} g^{\mu\mu} D_\mu D_\mu + \frac{m^2 c^2}{\hbar^2}\right) \Phi(\vec{r}, t) = 0. \tag{9.24}$$

After multiplying both sides by $e^{-ig\Lambda(\vec{r},t)}$, we recover the original equation (9.23). Therefore we talk about a *gauge symmetry*.

These computations seem complicated, but they are of fundamental relevance in particle physics, since they allow the following reasoning. First we postulate that the Klein–Gordon equation should be invariant under gauge transformations (9.14) in the above sense. Then we are forced to introduce a gauge field (a vector field) $A_\mu(\vec{r}, t)$, which has to transform as in (9.18) and has to be added in a particular way—inside the covariant derivative D_μ—to the modified Klein–Gordon equation (9.23). (These terms depending on $A_\mu(\vec{r}, t)$ are the first examples of coupling terms.) In short, postulating a gauge symmetry explains the existence of a gauge field. Since we can identify A_μ with the electromagnetic fields ϕ and A_i discussed in Chap. 5, the existence of the photon (and thus of light and all additional electromagnetic phenomena) follows from the postulate of a gauge symmetry.

Now the solutions of the new Klein–Gordon equation (9.23) for $\Phi(\vec{r}, t)$ depend on $A_\mu(\vec{r}, t)$ in D_μ. Then, no exact solutions can be found for $\Phi(\vec{r}, t)$ in general. Fortunately, however, we can assume in most cases that $A_\mu(\vec{r}, t)$ is negligibly small. Then the new Klein–Gordon equation (9.23) turns into the old Klein–Gordon equation (9.15), which has the solutions discussed in Chap. 4.

If $A_\mu(\vec{r}, t)$ (or, more precisely, $gA_\mu(\vec{r}, t)$) differs from zero but remains relatively small, we can solve (9.23) for $\Phi(\vec{r}, t)$ systematically in a power series in $gA_\mu(\vec{r}, t)$. This method requires the formalism of Green's functions, which we will not discuss here. Thereby it becomes clear, however, that the existence of coupling terms $gA_\mu(\vec{r}, t)$ in (9.23) implies that the field $\Phi(\vec{r}, t)$ can emit and absorb photons!

In Sect. 5.3 we described the emission and absorption of photons by vertices in Figs. 5.3 and 5.4. Connected to these vertices is a constant g, which depends on the electric charge q_e of the emitting or absorbing fields as in (5.26). In fact, the constant g is the same as the one appearing in the gauge transformations (9.14) and hence in the covariant derivative (9.21) in the term $gA_\mu(\vec{r}, t)$: the fact that $A_\mu(\vec{r}, t)$ in (9.23) is always multiplied by g explains why vertices are proportional to g.

Since g depends on the electric charge of the considered fields $\Phi(\vec{r}, t)$ as in (5.26) (and vanishes correspondingly for neutral fields), also the gauge transformations (9.14) of all fields—quarks and leptons such as the electron—depend on their electric charge. On the other hand the gauge transformation (9.18) of the photon field $A_\mu(\vec{r}, t)$, i.e., the function $\Lambda(\vec{r}, t)$, is the same always.

Up to now we have studied the gauge invariance of the Klein–Gordon equation for charged fields $\Phi(\vec{r}, t)$, but we also have to consider the corresponding equation for the photon field $A_\mu(\vec{r}, t)$. In (5.3) in Chap. 5 we assumed that the components $\phi(\vec{r}, t)$ and $\vec{A}(\vec{r}, t)$ of $A_\mu(\vec{r}, t)$ obey the Klein–Gordon equation as well. However, this claim is not quite correct: the original equation for the components of $A_\mu(\vec{r}, t)$ has the form

$$\sum_{\mu=0}^{3} g^{\mu\mu} \frac{\partial}{\partial x^\mu} \left(\frac{\partial}{\partial x^\mu} A_\sigma(\vec{r}, t) - \frac{\partial}{\partial x^\sigma} A_\mu(\vec{r}, t) \right) = 0. \tag{9.25}$$

(These are four equations for the four possible values of the index σ, whereas we sum over the indices μ.) Only this considerably more complicated version of the Klein–Gordon equation is invariant under the gauge transformations (9.18): the additional terms on the left-hand side

$$\sum_{\mu=0}^{3} g^{\mu\mu} \frac{\partial}{\partial x^\mu} \left(\frac{\partial}{\partial x^\mu} \frac{\partial}{\partial x^\sigma} \Lambda(\vec{r}, t) - \frac{\partial}{\partial x^\sigma} \frac{\partial}{\partial x^\mu} \Lambda(\vec{r}, t) \right) \tag{9.26}$$

cancel. However, (9.25) is somewhat awkward to solve since it mixes different components of $A_\mu(\vec{r}, t)$. Fortunately we can simplify life by using the freedom of a gauge transformation (9.18) with an arbitrary function $\Lambda(\vec{r}, t)$: we can always find a function $\Lambda(\vec{r}, t)$ depending on $A_\mu(\vec{r}, t)$ such that *after* a gauge transformation $A'_\mu(\vec{r}, t)$ satisfies the equation

$$\sum_{\mu=0}^{3} g^{\mu\mu} \frac{\partial}{\partial x^\mu} A'_\mu(\vec{r}, t) = 0. \tag{9.27}$$

If we assume that such a gauge transformation has been performed and $A_\mu(\vec{r}, t)$ satisfies (9.27) (this is the so-called Landau gauge), the second term in (9.25) drops out and (9.25) simplifies to

$$\sum_{\mu=0}^{3} g^{\mu\mu} \frac{\partial}{\partial x^\mu} \frac{\partial}{\partial x^\mu} A_\sigma(\vec{r}, t) = 0. \tag{9.28}$$

These are nothing but four (massless) Klein–Gordon equations of the kind (9.15), written in our new notation, for the four components $A_\sigma(\vec{r}, t)$ as used in Chap. 5, e.g., in (5.3).

Furthermore we observe that a mass term of the form $m^2 c^2/\hbar^2$, as present in (9.15), is forbidden in (9.25) and hence in (9.28): it would violate the invariance of

(9.25) under the gauge transformation (9.18). However, gauge fields (or, generally, vector fields) without a corresponding gauge symmetry of all corresponding equations would lead to inconsistencies, such as negative probabilities; thus violations of the corresponding gauge symmetries are absolutely taboo in every theory with gauge fields. This explains the assertion made in Sect. 7.3 that carriers of interactions—fields such as the photon—must be massless particles. (The generation of masses for gauge bosons by Higgs fields will be repeated below.)

Finally we can verify whether the relation between the components $A_0 = \phi/c$ and $\vec{A}$ of A_μ and the electric fields $\vec{E}$ and the magnetic fields $\vec{B}$ (see (5.4) and (5.5)) is invariant under gauge transformations. If we write the gauge transformation (9.18) separately for ϕ and $\vec{A}$, we obtain

$$\phi'(\vec{r}, t) = \phi(\vec{r}, t) + c\frac{\partial}{\partial x^0}\Lambda(\vec{r}, t) = \phi(\vec{r}, t) + \frac{\partial}{\partial t}\Lambda(\vec{r}, t),$$

$$A_i'(\vec{r}, t) = A_i(\vec{r}, t) + \frac{\partial}{\partial x^i}\Lambda(\vec{r}, t). \tag{9.29}$$

Replacing the fields ϕ and $\vec{A}$ in (5.4) and (5.5) by ϕ' and $\vec{A}'$, we find (after some calculation) that the derivative terms of $\Lambda(\vec{r}, t)$ on the right-hand side cancel and, accordingly, the fields $\vec{E}$ and $\vec{B}$ are *invariant* under a gauge transformation. This nice result formed the basis of the idea of gauge symmetry.

Our considerations so far of gauge symmetries were based on a generalization of the internal (symmetry) transformation (9.9) of a complex field, where the parameter Λ was replaced by $g\Lambda(\vec{r}, t)$. Such transformations are denoted as U(1) transformations, and the resulting gauge field theory as an *abelian* gauge theory.

The same considerations can be applied to internal transformations of the kind (9.13), where the matrix U_{ij} is either

- a 3×3 matrix rotating the various color components of the quark fields into each other (corresponding to a SU(3) transformation) or
- a 2×2 matrix rotating the various components of the quark and lepton isospin doublets into each other (corresponding to a SU(2) transformation).

Again we can postulate that the equations for the quark and lepton fields will be invariant under such transformations, even if the elements of the matrices U_{ij} are arbitrary functions depending on $\vec{r}$ and t (similar to $\Lambda(\vec{r}, t)$). Again we are "forced" to introduce additional gauge fields; such gauge theories are called *non-abelian* gauge theories or *Yang–Mills theories*, after C.N. Yang and R.L. Mills [31].

Invariance under the SU(3) transformations depending on $\vec{r}$ and t in the color space of quarks requires the introduction of eight gauge fields, which are the gluons introduced in Chap. 6. Now the gauge fields have to be 3×3 matrices in color space of the form A_μ^{ij}. (These matrices must satisfy $A_\mu^{ij} = A_\mu^{ji*}$ and must be traceless; there exist precisely eight independent such matrices, see the exercises at the end of this chapter.) Again they appear in equations for the quark fields in the form of the replacement of the ordinary derivative by a covariant derivative analogous to (9.17) above, where g has to be replaced by the coupling constant of the strong

interaction. Furthermore the gauge transformation of the gluon fields themselves is
more complicated than in (9.18) above for the photon field; now it is of the form

$$A_\mu^{ij}(\vec{r},t) \to A_\mu^{ij'}(\vec{r},t) = \sum_{k=1}^{3}\sum_{l=1}^{3} U_{ik}(\vec{r},t)A_\mu^{kl}(\vec{r},t)U_{jl}^*(\vec{r},t)$$

$$+ \frac{i}{g}\sum_{k=1}^{3} U_{ik}(\vec{r},t)\frac{\partial}{\partial x^\mu}U_{jk}^*(\vec{r},t). \tag{9.30}$$

(Omitting naively all indices and replacing the matrices U_{ij} by $e^{ig\Lambda(\vec{r},t)}$, we re-obtain
the transformation law (9.18).)

Also the equation for the gluon fields corresponding to Eq. (9.25) above is considerably more complicated, and contains additional terms quadratic and trilinear in the
fields A_μ^{ij} such that it is invariant under the gauge transformations (9.30). These terms
are responsible for the three- and four-gluon vertices in Figs. 6.5 and 6.6.

However, the essential point is that the postulate of an invariance of the fundamental equations under SU(3) transformations in color space alone suffices to give
reasons for the existence of gluon fields and hence the strong interaction and quantum
chromodynamics [32, 33].

This reasoning is also valid for the weak interaction: invariance under SU(2)
transformations depending on $\vec{r}$ and t in isospin space of quarks and leptons requires
the introduction of three more gauge fields (2×2 matrices in isospin space). Two of
these can be identified with the known $W^\pm$ bosons.

In order to understand the origin of the $W^\pm$ boson masses by the Higgs field, we
first have to take into account additional terms in the Eqs. (9.25) and (9.28) for the
gauge fields. Similar to how gauge fields appear in the (modified) equations of the
kind (9.23) for fields Φ, the fields Φ appear in the equations for the gauge fields. Up
to now we have omitted such terms in Eqs. (9.25) and (9.28), since we can usually
assume that such fields Φ are negligibly small.

An exception is the Higgs field, for which we assumed in Sect. 7.3 that it takes
on a non-vanishing constant value everywhere in the universe. Subsequently we
will investigate the influence of a scalar field Φ on the equation of the form (9.28)
for a simple gauge field, such as the photon; the result can also be applied to the
corresponding (but more complicated) equation for the $W^\pm$ bosons.

In order to allow for a scalar field $\Phi(\vec{r},t)$ with coupling constant g to a gauge
field A_σ, (9.28) is modified as follows:

$$\sum_{\mu=0}^{3} g^{\mu\mu}\frac{\partial}{\partial x^\mu}\frac{\partial}{\partial x^\mu}A_\sigma + \frac{ig}{\hbar^2 c^2}\left[\left(\Phi^*\frac{\partial}{\partial x^\sigma}\Phi - \Phi\frac{\partial}{\partial x^\sigma}\Phi^*\right) - 2igA_\sigma\Phi^*\Phi\right] = 0. \tag{9.31}$$

First we can verify that the additional terms in the square brackets are invariant under
the gauge transformations (9.14) and (9.18) by themselves—in contrast to a mass
term for A_σ. Next we assume that Φ stands for a Higgs field assuming a constant

value H (with H real, i.e., $H^* = H$) everywhere in the universe. Then the derivative terms of Φ vanish, and (9.31) simplifies to

$$\sum_{\mu=0}^{3} g^{\mu\mu} \frac{\partial}{\partial x^\mu} \frac{\partial}{\partial x^\mu} A_\sigma(\vec{r}, t) + 2\frac{g^2}{\hbar^2 c^2} H^2 A_\sigma = 0. \tag{9.32}$$

This equation is of the same form as the massive Klein–Gordon equation (9.15) (which was nothing but a rephrasing of the Klein–Gordon equation (4.11)), if we identify $2(g^2/c^2)H^2$ in (9.32) with $m^2 c^2$ in (9.15). Again we see—as in Sect. 7.3— how a constant Higgs field generates a mass, but in contrast to Sect. 7.3 we have used the language of field theory (i.e., the Klein–Gordon equation). However, the mass generation for gauge fields occurs only if the Higgs field is present in the corresponding equation. This is not the case for photons (since the Higgs field is neutral) nor for gluons (since the Higgs field carries no color)—in these cases the coupling constant g in the above equations vanishes. In the case of $W^\pm$, we just have to replace the coupling g_w of the weak interaction for g and correct a factor of 2 (because of the SU(2) index structure). Then we obtain (7.17) for the $W^\pm$ mass depending on the value H of the Higgs field.

One complication is, however, that the complete theory of electromagnetic and weak interactions is based on a SU(2) *and* a U(1)$_Y$ gauge symmetry, where the latter is distinct from the pure electromagnetic gauge symmetry: the charges of the quarks and leptons with respect to U(1)$_Y$ gauge transformations differ from their electric charges, and the gauge field B_μ corresponding to the U(1)$_Y$ gauge symmetry is not identical to the photon. Once the Higgs field H assumes a constant value H, we find that

- the $W_\mu^\pm$ bosons become massive as stated above,
- the third gauge boson W_μ^3 of the SU(2) gauge symmetry mixes with the B_μ gauge boson of the U(1)$_Y$ gauge symmetry (they have to be rotated into each other). One linear combination corresponds to the known massless photon A_μ, and the other linear combination to the massive Z_μ boson mentioned in Chap. 7:

$$A_\mu = \cos\theta_w\, B_\mu + \sin\theta_w\, W_\mu^3$$
$$Z_\mu = \cos\theta_w\, W_\mu^3 - \sin\theta_w\, B_\mu. \tag{9.33}$$

If we denote the coupling constant of the SU(2) gauge symmetry by g_2 and the coupling constant of the U(1)$_Y$ gauge symmetry by g_1, the "weak mixing angle" θ_w is given by

$$\sin\theta_w = \frac{g_1}{\sqrt{g_1^2 + g_2^2}}. \tag{9.34}$$

For the electric elementary charge we find

$$e = g_1 \cos\theta_w = g_2 \sin\theta_w, \tag{9.35}$$

which corresponds to the relation (7.5) between α and α_w, since α_w is proportional to g_2^2. The cosine of the weak mixing angle $\cos\theta_w$ appears also in the ratio of the couplings of the Z boson and the $W^\pm$ boson to the Higgs boson and hence, as in (7.12), in the ratio of the Z boson mass to the W boson mass. The 1979 Nobel prize was awarded to S.L. Glashow, A. Salam, and S. Weinberg for the development of this relatively complicated theory of the weak and electromagnetic interactions [34–37].

Apart from this complication, we have now traced back also the third—weak—interaction to a gauge symmetry. However, this gauge symmetry is spontaneously broken by the constant value H of the Higgs field, which explains the distinct properties of the components of the isospin doublets (quarks and leptons) as well as the masses of the W and Z bosons.

Finally the fourth interaction (gravity) can also be based on a symmetry of equations like the Klein–Gordon equation (9.15). However, now the corresponding transformations are not gauge transformations of fields like Φ or A_μ, but so-called general coordinate transformations: this means nothing but a change of the choice of coordinates, such as the transition from rectangular coordinates x and y in the plane to polar coordinates ρ and θ, or the transition from rectangular coordinates x, y, and z in space to spherical coordinates ρ, θ, and ϕ. Such coordinate transformations can always be described as a replacement of the original coordinates by functions of the new coordinates; in the plane, e.g., by

$$x \to x(\rho,\theta) = \rho\cos\theta,$$
$$y \to y(\rho,\theta) = \rho\sin\theta, \qquad (9.36)$$

considering subsequently the new coordinates as independent variables.

In a four-dimensional space-time, general coordinate transformations are of the form

$$x^\mu \to x^\mu(x'^\nu), \qquad (9.37)$$

where $x^\mu(x'^\nu)$ can be four arbitrary functions of the four new coordinates x'^ν, and subsequently x'^ν are considered as independent variables.

Accordingly derivatives with respect to x^μ have to be rewritten in terms of derivatives with respect to x'^ν according to the chain rule:

$$\frac{\partial}{\partial x^\mu} = \sum_{\nu=0}^{3} \frac{\partial x'^\nu}{\partial x^\mu} \frac{\partial}{\partial x'^\nu}. \qquad (9.38)$$

Let us take another look at the Klein–Gordon equation (9.15). After a general coordinate transformation—and the corresponding required replacement of derivatives as in (9.38)—it will become considerably more complicated. This can be avoided if we replace the Minkowski metric $g^{\mu\nu}$ in (9.15) by a *field* $g^{\mu\nu}(x)$ and replace, after a general coordinate transformation, $g^{\mu\nu}(x)$ by $g'^{\mu\nu}(x')$ with

$$g'^{\mu\nu}(x') = \sum_{\rho=0}^{3}\sum_{\sigma=0}^{3} \frac{\partial x'^\mu}{\partial x^\rho} \frac{\partial x'^\nu}{\partial x^\sigma} g^{\rho\sigma}(x(x')). \qquad (9.39)$$

In Sect. 3.2 we treated the metric $g^{\mu\nu}$ as a field depending, in general, on $\vec{r}$ and t; however, at that time we did not consider general coordinate transformations. (Expressions for g_{00} as in (3.32) are only valid in a given coordinate system.) The necessity of transforming the metric under coordinate transformations as in (9.39) follows, however, from the expression (3.32) for the distance Δ^{AB}, which must *not* depend on the choice of coordinates.

If we replace the Klein–Gordon equation (9.15) by

$$\left(\sum_{\mu=0}^{3}\sum_{\nu=0}^{3}\frac{\partial}{\partial x^{\mu}}\sqrt{-g}\,g^{\mu\nu}\frac{\partial}{\partial x^{\nu}}+\sqrt{-g}\frac{m^{2}c^{2}}{\hbar^{2}}\right)\Phi(\vec{r},t)=0, \qquad (9.40)$$

where g is the determinant of $g_{\mu\nu}$, we can show that the equation is invariant under general coordinate transformations. (For a constant Minkowski metric we re-obtain from (9.40) the original equation (9.15).)

Here we will not require a complete understanding of the technical details; we only want to point out that another postulate of a symmetry of equations—here under general coordinate transformations—implies again the introduction of a "gauge field" (here: the metric), which implies another interaction (here: gravity). The development of the theory of general relativity on the basis of this postulate was one of the best ideas (or *the* best idea) of Albert Einstein.

To summarize, we have now traced back *all* four fundamental interactions to symmetries of equations—for this reason symmetries play such an important role in fundamental physics.

Exercises (Challenging!)

9.1. Consider an internal transformation (9.13) generated by complex $N \times N$ matrices U_{ij}. In the case of "small" transformations, these matrices can be written as $U_{ij} = \delta_{ij} + iA_{ij} + \dots$ with $|A_{ij}| \ll 1$.

(a) Deduce, with the help of a series expansion to first order in A_{ij}, the hermiticity of A_{ij} (i.e., $A_{ij} = A_{ji}^{*}$) from the unitarity of U_{ij} (i.e., $\sum_{j=1}^{N} U_{ij}U_{kj}^{*} = \delta_{ik}$).

(b) Deduce from $det(U) = 1$ the vanishing of the trace of A_{ij} (i.e., $\sum_{i=1}^{N} A_{ii} = 0$).

9.2. Deduce the number of linearly independent matrices A_{ij} that are Hermitian and traceless for $N = 2$ and $N = 3$. (This number is equal to the number of gauge bosons of a SU(N) gauge theory.)

Chapter 10
The Standard Model of Particle Physics

The current version of the Standard Model of particle physics is based on only a few elementary particles: six quarks and six leptons, the gauge fields responsible for the interactions, and the still sought-after Higgs boson. The fundamental interactions relevant for particle physics are the electromagnetic, strong, and weak interactions. This chapter summarizes the properties of the elementary particles, the aspects shared by the fundamental interactions and where they differ, and the open questions left unanswered by the Standard Model.

10.1 Properties of the Elementary Particles

In Chaps. 5, 6 and 7 we discussed the three fundamental interactions (or forces): electromagnetism and the strong and weak interactions. In principle, gravity is a fourth fundamental interaction, however, it plays practically no role in particle physics and is neglected in the so-called "Standard Model".

We can divide the elementary particles into

- "matter" (the quarks and leptons with spin $\hbar/2$);
- bosons with spin $\hbar$, whose exchange generates the interactions (or forces), and the Higgs boson with spin zero.

Up to now six quarks and six leptons have been discovered, see Chaps. 6 and 7. Table 10.1 summarizes their properties again (where the electric charge is given in multiples of e, and the light quark masses correspond to those in potential models); the properties of the bosons known nowadays are given in Table 10.2.

U. Ellwanger, *From the Universe to the Elementary Particles*,
Undergraduate Lecture Notes in Physics, DOI: 10.1007/978-3-642-24375-2_10,
© Springer-Verlag Berlin Heidelberg 2012

Table 10.1 Masses, electric charges, and interactions of the known quarks and leptons

	Mass	Charge	Strong Int.	Weak Int.
Quark				
u	~ 300 MeV/c^2	$+2/3$	Yes	Yes
d	~ 300 MeV/c^2	$-1/3$	Yes	Yes
s	~ 500 MeV/c^2	$-1/3$	Yes	Yes
c	~ 1.4 GeV/c^2	$+2/3$	Yes	Yes
b	~ 4.4 GeV/c^2	$-1/3$	Yes	Yes
t	~ 173 GeV/c^2	$+2/3$	Yes	Yes
Lepton				
ν_e	< 2 eV/c^2	0	No	Yes
e^-	~ 0.511 MeV/c^2	-1	No	Yes
ν_μ	<190 keV/c^2	0	No	Yes
μ^-	~ 106 MeV/c^2	-1	No	Yes
ν_τ	<18 MeV/c^2	0	No	Yes
τ^-	~ 1.78 GeV/c^2	-1	No	Yes

Table 10.2 Masses, electric charges, and interactions of the known bosons

Boson	Mass (GeV/c^2)	Charge	Strong Int.	Weak Int.
Photon (γ)	0	0	No	No
Gluon	0	0	Yes	No
$W^\pm$	~ 80.4	± 1	No	Yes
Z	~ 91.2	0	No	Yes
Higgs	$\gtrsim 114$	0	No	Yes

10.2 Properties of the Fundamental Interactions

The essential properties of the three interactions of particle physics are as follows:

- The electromagnetic interaction is generated by the exchange of massless photons, which carry no electric charge by themselves. The value of the electric fine structure constant $\alpha \sim 1/137$ is relatively small, implying that the description of electromagnetic processes in quantum field theory is relatively simple: in most cases it suffices to confine oneself to the simplest Feynman diagrams contributing to a given process.
- The strong interaction is generated by the exchange of massless gluons. The quantity corresponding to the electric charge is color, which is carried by quarks but not by leptons. Gluons themselves carry color, i.e., a strong charge, as well. The value of the strong fine structure constant is $\alpha_s \sim 1$, and accordingly we have to take into account all Feynman diagrams contributing to a given process. The most important consequence is that the dependence of the strong force on the distance between colored particles is very different from that for the electric force: at large distances, the strong force remains constant (and attractive), leading to the confinement of

quarks and gluons inside hadrons. Hadrons are either baryons, consisting of three quarks, or mesons, consisting of a quark and an antiquark.

- The weak interaction is generated by the exchange of $W^\pm$ and Z^0 bosons. These bosons are (very) massive, implying that this interaction is relatively weak. The explanation of these masses necessitates the introduction of the Higgs field, the non-vanishing constant value of which everywhere generates an effective mass for every particle coupling to the Higgs boson. The masses of all elementary particles—including quarks and leptons—are generated this way.

To date (September 2011), the existence of the Higgs boson has not been confirmed, and its mass is still unknown.

What are the "fundamental" (not calculable) parameters of the standard model? First, these are the three fine structure constants of the electromagnetic, strong, and weak interactions. To these we have to add the six quark masses or, alternatively, the six corresponding Yukawa couplings (see (7.20)). Since the three different quark families (7.1) can transform into each other during processes of the weak interaction (see Fig. 7.2), they are "rotated" into each other (similarly to the neutrinos in (7.26) treated—simplified—in Sect. 7.5), which leads to three real and one imaginary mixing angle or elements of the Cabibbo–Kobayashi–Maskawa matrix. (The imaginary mixing angle, implied by a complex mass parameter or Yukawa coupling, respectively, allows a description of CP violation, mentioned in Sect. 7.4.) Thus the quark masses and mixing angles alone lead to 10 additional parameters. First of all, the masses of the three charged leptons correspond to three more parameters. However, the phenomenon of neutrino oscillations indicates that the complete lepton sector involves at least 10 parameters as well—possibly even more. Finally, the expression (7.16) for the potential energy of the Higgs field contains two additional parameters: μ^2 and λ_H^2. Adding gravity to the fundamental interactions gives two more parameters: Newton's constant G and the cosmological constant Λ.

10.3 Open Questions

The Standard Model defined in terms of the particles in Tables 10.1 and 10.2 and the three fundamental interactions describes successfully a very large number of processes; it is not in conflict with any of the numerous results of measurements. However, several questions remain open:

(a) Why are there three families of quarks and leptons with nearly identical properties? They differ only in their masses, i.e., (Yukawa) couplings to the Higgs field; why are these couplings so different? What is the origin of the mixing angles?

(b) What is the structure of the neutrino mass terms (see Sect. 7.5)? Do right-handed neutrinos exist?

(c) Why are there three interactions and what is the origin of the values of their couplings, i.e., fine structure constants? (A possible answer to this question is the theory of "Grand Unification" considered in Sect. 12.1.)

(d) If we take quantum corrections in quantum field theory into account, a numerical conflict related to the parameter μ in (7.16) for the potential energy of the Higgs field appears. This numerical conflict is similar to the problem of the cosmological constant mentioned at the end of Sect. 7.3 and will be discussed in Sect. 12.2 together with a possible solution (supersymmetry).

(e) If we describe gravity in quantum field theory, quantum corrections lead to infinite results (see Sect. 12.3). A possible solution of this fundamental conflict between quantum field theory and the theory of general relativity is string theory.

Chapter 11
Quantum Corrections and the Renormalization Group Equations

In the framework of quantum field theory, measured coupling constants (which characterize the different fundamental interactions) depend on the energy at which the experiments are carried out. Furthermore one has to distinguish between measured and fundamental coupling constants. These concepts coincide with energy-dependent measurements of the strong coupling constant.

11.1 Quantum Corrections

In this chapter we will deduce the surprising result that coupling constants depend on the energy of the particles participating in the measurement process.

We recall how the probability of a process has to be computed in quantum field theory:

(a) First we have to find all Feynman diagrams that contribute, in principle, to the process under consideration. (In the case of small coupling constants, only the simplest diagrams give numerically relevant contributions.)

(b) The contribution of every diagram to the amplitude (whose square gives the probability of a given process) is to be computed with help of the Feynman rules, see Chap. 5.

In the case of electron–electron scattering, the simplest diagram is of the form in Fig. 11.1.

With help of the vertices given in Chap. 5 we can also draw the diagram in Fig. 11.2, which is a diagram with four vertices. This diagram describes (amongst others) the following process: at vertex A an electron emits a photon. At vertex B this photon decays into an electron–positron pair. After a "loop" this pair annihilates at vertex C into another photon, which is absorbed by the other electron at vertex D.

We can count the powers of $\hbar$ originating from the vertices, the propagators, and the remaining factors. Compared to the diagram in Fig. 11.1 with two vertices (where all powers of $\hbar$ cancel), all diagrams with four vertices contain an additional power

U. Ellwanger, *From the Universe to the Elementary Particles*,
Undergraduate Lecture Notes in Physics, DOI: 10.1007/978-3-642-24375-2_11,
© Springer-Verlag Berlin Heidelberg 2012

Fig. 11.1 The simplest diagram contributing to electron–electron scattering

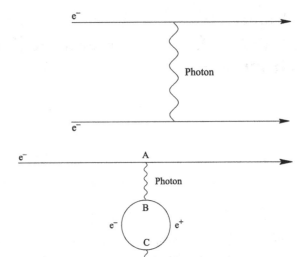

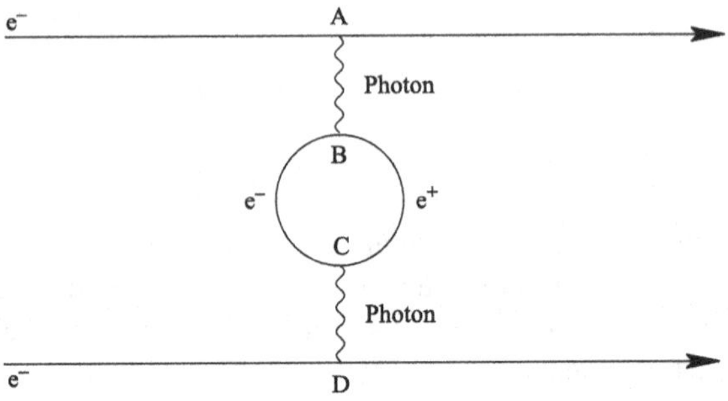

Fig. 11.2 A loop diagram contributing to electron–electron scattering

of $\hbar$. An additional power of $\hbar$ means that we are dealing with a quantum effect, which would vanish in classical electrodynamics obtained in the limit $\hbar \to 0$. For this reason we denote these contributions as quantum corrections. We should note that there exist more diagrams with four vertices, which have to be treated in a similar way as the diagram in Fig. 11.2 in the following.

Next we study the consequences of energy conservation for the energy of each of the particles; in Fig. 11.2 the photons and the electron–positron pair in the loop are virtual particles. First we find for the electrons in the initial and final states, exactly as in Fig. 11.1 and as discussed in Chap. 5, that all their energies are the same if the momenta before scattering are directed in opposite directions (which we will assume in the following). It follows that the energies of the virtual photons between A and B and also C and D vanish.

Concerning the energies of the electron and the positron in the loop, we have to emphasize that energies of virtual particles can also be negative. Thus energy conservation at the vertices B and C implies that the energies of the electron and the positron are of the same modulus but of opposite sign, such that their sum vanishes. However, the moduli of these energies, denoted by Q subsequently, are *not* determined by energy conservation!

Which value for Q should we thus chose? The fundamental rule of quantum mechanics tells us that we have to sum over all the processes that are allowed by the conservation laws. Hence we have to sum, that is, integrate, over all values of Q between $-\infty$ and $+\infty$.

Taking the dependence of the propagators of the electron–positron pair on the energy and hence on Q into account, we find that the contribution of the diagram in

Fig. 11.2 to the amplitude contains an integral of the form

$$\int_0^\infty dQ^2 \frac{1}{Q^2 + m_e^2} \tag{11.1}$$

where m_e is the electron (and positron) mass, and we have introduced the integration variable Q^2, which is always positive. (In principle we also have to integrate over the momenta of the electron and the positron; as the final result we find indeed an integral of the form (11.1). Actually we would have to replace m_e by $m_e c^2$ in the denominator of (11.1); for simplicity we omit these powers of c in the following.)

Let us try to evaluate this integral. First we have to find a function $F(Q^2)$ whose derivative with respect to Q^2 gives the expression under the integral. Up to an irrelevant constant, this function is given in our case by

$$F(Q^2) = \ln(Q^2 + m_e^2). \tag{11.2}$$

Then the integral (11.1) is obtained by this function with Q^2 replaced by the upper limit of the integral less the function evaluated at the lower limit of the integral. Whereas the evaluation of the function (11.2) at the lower limit $Q^2 = 0$ of the integral poses no problem (we obtain $\ln(m_e^2)$), we find for the function at the upper limit $Q^2 = \infty$ also $+\infty$, since the logarithm of $+\infty$ is infinite as well – hence no reasonable result. In the early days of quantum field theory this was considered a serious problem.

Now we treat this problem as follows: instead of integrating over all possible values of Q^2 from 0 to $+\infty$, we integrate only over a finite interval between 0 and Λ^2. The quantity Λ^2 has the following meaning: the expression under the integral (11.1) was deduced under the assumption that we know the energy dependence of the propagators (and all other factors contributing to the calculation). However, we can be sure about this energy dependence only for energies that have been experimentally verified. In fact we can assume that the used Feynman rules are valid up to an upper bound $|Q| \leq \Lambda$, but the larger the chosen value of Λ the larger the underlying assumption. The agreement between high-energy experiments (employing particles with energies up to about 1 TeV = 1000 GeV) and theory (assuming standard Feynman rules) allows us to conclude that we can assume $\Lambda \gtrsim 1$ TeV.

As a result of this reasoning we replace the upper limit of the integral (11.1) by Λ^2. Now we obtain a finite result for this integral, but the result depends on Λ:

$$\int_0^{\Lambda^2} dQ^2 \frac{1}{Q^2 + m_e^2} = \ln\left(\frac{\Lambda^2 + m_e^2}{m_e^2}\right) \simeq \ln\left(\frac{\Lambda^2}{m_e^2}\right), \tag{11.3}$$

where we have made the assumption $\Lambda \gg m_e$.

What does this imply for the contribution of the diagram in Fig. 11.2 to the amplitude of the considered process? Let us compare again the powers of the coupling constant g, i.e., the fine structure constant $\alpha = g^2 \hbar/(4\pi)$ of the contributions of the diagrams in Figs. 11.1 and 11.2 to the amplitude A: the contribution of the diagram in Fig. 11.1 is of the form $\alpha A^{(1)}(\theta)$, since it contains two vertices. The

Fig. 11.3 Multi-loop
diagrams contributing to
electron–electron scattering

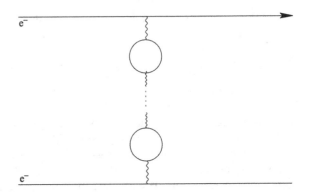

contribution of the diagram in Fig. 11.2 is of the form $\alpha^2 \left(\ln \left(\Lambda^2/m_{\mathrm{e}}^2 \right) A^{(2)}(\theta) + \ldots \right)$,
where we have written explicitly the dependence on Λ, and the dots denote terms
independent of Λ or negligible for $\Lambda \gg m_{\mathrm{e}}$. The dependence of $A^{(1)}(\theta)$ and $A^{(2)}(\theta)$
on the momenta of the external electrons can be computed; from (5.28) we obtain,
with (5.21) for $\left| \vec{p}^{\,\mathrm{ph}} \right|$, $A^{(1)}(\theta) = 8\pi m_{\mathrm{e}}^2 \hbar c/[|\vec{p}^{\,a}|^2 (1 - \cos\theta)]$ since we bracketed off
a factor α.

We explained in Chap. 5 that the contributions of the diagrams with four vertices
are relatively small owing to the small value of α. Now we see that this conclusion was
not necessarily justified: if the logarithm multiplying $A^{(2)}$ is very large (if $\Lambda \gg m_{\mathrm{e}}$
holds), the contribution of the diagram in Fig. 11.2 can be even larger than the
contribution of the diagram in Fig. 11.1! (The original conclusion remains valid only
for the terms indicated above by dots, leaving aside the large logarithm.)

In fact (and luckily) we find that the dependence of $A^{(2)}(\theta)$ on the scattering angle
θ, that is, the momenta of the external electrons, coincides with that of $A^{(1)}(\theta)$ up
to a constant $-b$. (The negative sign in front of b is a convention.) This allows us to
combine the numerically dominant contributions as follows:

$$A(\theta) \simeq \alpha A^{(1)}(\theta) + \alpha^2 \ln \left(\frac{\Lambda^2}{m_{\mathrm{e}}^2} \right) A^{(2)}(\theta) = \alpha A^{(1)}(\theta) \left(1 - b\alpha \, \ln \left(\frac{\Lambda^2}{m_{\mathrm{e}}^2} \right) \right).$$
(11.4)

This is not the end of the story: there exist additional diagrams with more loops
of electron–positron pairs as in Fig. 11.3.

Each loop in a diagram corresponds to another power of α, and implies an addi-
tional power of the logarithm $\ln \left(\Lambda^2/m_{\mathrm{e}}^2 \right)$ as well as the constant b in the contribution
of the corresponding diagram to the amplitude. We can evaluate the sum over all these
diagrams (a geometric series) and find that (11.4) has to be replaced by

$$A(\theta) = \frac{\alpha A^{(1)}(\theta)}{1 + b\alpha \, \ln \left(\Lambda^2/m_{\mathrm{e}}^2 \right)}.$$
(11.5)

(We can verify that (11.5) coincides with (11.4) in a power series expansion in α to
the order α^2.)

11.2 Energy Dependent Coupling Constants

Now we assume that we want to use electron–electron scattering in order to measure the fine structure constant α. To this end we proceed as follows: First we measure the number of scattered electrons as a function of the scattering angle θ, or rather the probability $P(\theta)$. $P(\theta)$ is proportional to the square of $A(\theta)$, see (5.29), which allows us to deduce $A(\theta)$ from the measurement. Finally we compare the result for $A(\theta)$ with the formula

$$A(\theta) = \alpha_{\text{measured}} A^{(1)}(\theta), \tag{11.6}$$

where the known expression above for $A^{(1)}(\theta)$ is used. We emphasize that the measurement cannot distinguish the contributions of the various diagrams; for this reason (11.6) is the only reasonable definition of the measured fine structure constant. In practice we use processes in atomic physics for the most precise measurements of α_{measured}, but also here we sum automatically over the corresponding Feynman diagrams.

Comparing the expressions (11.5) and (11.6) we find

$$\alpha_{\text{measured}} = \frac{\alpha}{1 + b\,\alpha\,\ln\left(\Lambda^2/m_{\text{e}}^2\right)}. \tag{11.7}$$

Accordingly we have to distinguish the "fundamental" coupling constant g (or rather $\alpha = g^2\hbar/(4\pi)$) from α_{measured}! The "fundamental" coupling constant g is the one connected to the vertices, i.e., the emission or absorption of a photon. We could measure this fundamental coupling only if we could determine the contribution of the diagram in Fig. 11.1 separately from the contributions of the diagrams in Figs. 11.2 and 11.3—this is impossible, however, and leads to (11.7) for α_{measured} as a function of α and Λ.

We should note that the expression (11.6) contains—by definition—the measurable quantities α_{measured} and the θ dependence of $A^{(1)}$ but no explicit dependence on Λ if it is expressed in terms of α_{measured}. This is possible since the θ dependence of the diagrams with loops (denoted as $A^{(2)}(\theta)$ above) coincides, up to a constant, with the θ dependence of $A^{(1)}(\theta)$. A theory in which all relations between measurable quantities are independent of Λ is denoted as *renormalizable*: in a renormalizable theory, Λ can in principle be arbitrarily large (even infinite). (The 1965 Nobel prize was awarded to S.-I. Tomonaga, J. Schwinger, and R.F. Feynman for, among other things, the proof that quantum electrodynamics is renormalizable in this sense.) In Chap. 12 we will see that this does not hold for quantum gravity.

Let us return to α_{measured}, whose value is known: $\alpha_{\text{measured}} \sim 1/137$. This value was measured in processes at vanishing energies of the photons exchanged in Figs. 11.1, 11.2, 11.3. We can imagine, however, that α_{measured} is measured in a process where the energies of the exchanged photons are no longer small. A typical example is the pair production of particles p and $\bar{\text{p}}$ described in Chap. 8 in Fig. 8.5. Here the energy E of the photon is given by $E = E(\text{e}^+) + E(\text{e}^-)$, which is typically much larger than the electron mass (multiplied by c^2).

Fig. 11.4 Loop diagram contributing to particle–antiparticle production via electron–electron annihilation

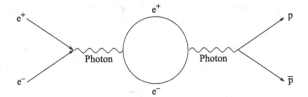

Loop diagrams as in Fig. 11.2 also contribute to this process, but they are now of the form of Fig. 11.4 and of the rotated Fig. 11.3.

Now we can repeat the previous considerations, starting with the integral over the energies of the particles e^+ and e^- in the loop. In contrast to above, the sum of these energies no longer vanishes but is equal to the non-vanishing photon energy E, owing to energy conservation at the vertices. (Still, the difference Q between these energies is undetermined.) This modifies the propagators of the particles in the loop, and as a result we find that the integral (11.3) over the undetermined energy difference Q assumes (approximately) the form

$$\int_0^{\Lambda^2} dQ^2 \frac{1}{Q^2 + E^2 + m_e^2}. \tag{11.8}$$

All further considerations remain unchanged: as the result for the integral we find the logarithm in (11.3) (with m_e^2 replaced by $E^2 + m_e^2$) and the θ dependences of the amplitudes corresponding to Figs. 8.5 and 11.4 are again the same; hence the measured fine structure constant α_{measured} can again be written as in (11.7)—with m_e^2 replaced by $E^2 + m_e^2$, however. Then we obtain in the case $E^2 \gg m_e^2$

$$\alpha_{\text{measured}} = \frac{\alpha}{1 + b\,\alpha\,\ln\left(\Lambda^2/E^2\right)}. \tag{11.9}$$

The important consequence of this equation is that, in the case of a measurement of a fine structure constant at an energy $E \neq 0$, the result of the measurement should depend on the energy! The reason for this energy dependence is the energy dependence of the contributions of the loop diagrams to the amplitudes.

Apart from the energy E, the following quantities appear in (11.9): the fundamental fine structure constant α, the parameter Λ, and the constant b. Whereas the fundamental fine structure constant α and the parameter Λ are unknown quantities, the constant b is calculable.

We can show, however, that the energy dependence of α_{measured} (we should write $\alpha_{\text{measured}}(E)$) can be given in terms of known quantities only. To this end we differentiate both sides of (11.9) with respect to $\ln\left(E^2\right)$ (using $\ln\left(\Lambda^2/E^2\right) = \ln\left(\Lambda^2\right) - \ln\left(E^2\right)$) and use (11.9) again on the right-hand side. The result can simply be written as

$$\frac{d\alpha_{\text{measured}}(E)}{d\ln\left(E^2\right)} = b\,\alpha_{\text{measured}}^2(E), \tag{11.10}$$

which is denoted as the *renormalization group equation* [38, 39]. This means that the variation of $\alpha_{measured}(E)$ with E (or rather $\ln\left(E^2\right)$) can be expressed in terms of $\alpha_{measured}(E)$ itself and the calculable constant b: if b is positive, $\alpha_{measured}(E)$ should increase with E, whereas $\alpha_{measured}(E)$ should decrease with E for b negative. In quantum electrodynamics one obtains a positive value for b.

We should note that this result is not quite complete: additional (more complicated) loop diagrams, which also contribute to the physical processes, imply a somewhat more complicated relation between the measured and the fundamental fine structure constant than in (11.7) or (11.9). This leads to additional terms in the calculation of the derivative of $\alpha_{measured}(E)$ with respect to $\ln\left(E^2\right)$; these additional terms can all be described by additional higher powers of $\alpha_{measured}(E)$ on the right-hand side of the renormalization group equation (11.10). Then we obtain (omitting the argument E of $\alpha_{measured}(E)$)

$$\frac{d\alpha_{measured}}{d\ln\left(E^2\right)} = b_1\,\alpha_{measured}^2 + b_2\,\alpha_{measured}^3 + \cdots \equiv \beta(\alpha_{measured}), \qquad (11.11)$$

where the function $\beta(\alpha_{measured})$ is generally denoted as the β function. Usually $\alpha_{measured}$ is not very large; then the decrease or increase of $\alpha_{measured}$ with E follows solely from the sign of the first term $\sim b_1$ ($= b$ in (11.10)) of the β function.

The same considerations can be applied to the strong and the weak fine structure constants. In the case of the strong interaction we have to replace the electrons and positrons in the diagrams in Figs. 11.1, 11.2, 11.3 and 11.4 by quarks and antiquarks, and the photons by gluons. (As a result of the vertices in Figs. 6.5 and 6.6, gluons in the loops contribute as well.)

Correspondingly we obtain again a difference between the measured strong fine structure constant $\alpha_{s,\,measured}$ and the "fundamental" fine structure constant α_s, which is again of the form (11.9):

$$\alpha_{s,\,measured} = \frac{\alpha_s}{1 + b_s\,\alpha_s\,\ln\left(\Lambda^2/E^2\right)}. \qquad (11.12)$$

In the case of the weak interaction, the photons in the diagrams 11.1–11.4 have to be replaced by $W^\pm$ and Z bosons, and the electrons or positrons in the loops by the sum of all particles (quarks, leptons, and the bosons themselves) that couple to $W^\pm$ and Z bosons. The relation between the measured weak fine structure constant $\alpha_{w,\,measured}$ and the "fundamental" weak fine structure constant α_w is again of the form (11.9):

$$\alpha_{w,\,measured} = \frac{\alpha_w}{1 + b_w\,\alpha_w\,\ln\left(\Lambda^2/E^2\right)}. \qquad (11.13)$$

The constants b_s and b_w can be computed; they depend on the number and couplings of the particles appearing in the loops. For energies E larger than all known particle masses we find

$$b_s = -\frac{7}{4\pi}, \qquad b_w = -\frac{19}{24\pi}. \tag{11.14}$$

We note that the same constants appear in the corresponding renormalization group equations (11.10) for $\alpha_{s,\,measured}$ and $\alpha_{w,\,measured}$. Here we made the assumption that the energy E is larger than all masses (multiplied by c^2) of the particles in the loops. If this assumption is not satisfied, the diagrams with particles with masses larger than E/c^2 do not contribute to the calculation of the constant b in (11.10), therefore these "constants" have little jumps at the corresponding energy: for an energy E where only n_f quarks are lighter than E/c^2, b_s in (11.10) for $\alpha_{s,\,measured}$ has to be replaced by

$$b_s = -\frac{1}{4\pi}\left(11 - \frac{2}{3}n_f\right); \tag{11.15}$$

for $E/c^2 > m_{top}$, where $n_f = 6$ holds, we recover the value in (11.14) for b_s from (11.15).

Since here these constants b_s and b_w are negative, the measured couplings, i.e., fine structure constants $\alpha_{s,\,measured}$ and $\alpha_{w,\,measured}$, should decrease with increasing energy, or increase with decreasing energy. This effect is the more important the larger the couplings themselves are—it should be most important for the coupling $\alpha_{s,measured}$ of the strong interaction. (The 2004 Nobel prize was awarded to D.J. Gross and F. Wilczek [40] and H.D. Politzer [41] for the discovery of this phenomenon.)

The coupling α_s has been measured in various processes at various energies E. (We omit the index "measured" from now on.) Some results of different experiments in the interval 22 GeV $< E <$ 189 GeV are summarized in Fig. 11.5, where the formula (11.12) with $b_s = -23/(12\pi)$ (from (11.15) with $n_f = 5$, since the top quark does not contribute for $E \lesssim m_{top} c^2$) should be approximately valid. The energies E are plotted along the horizontal axis, and the results of measurements of α_s together with their error bars along the vertical axis. The data for $E \simeq 22, 35,$ and 44 GeV [42,43], $E \simeq 58$ GeV [44], $E \simeq 133$ GeV [45, 46], and $E \simeq 189$ GeV [47–50] have been obtained from $e^+ + e^- \rightarrow$ hadrons, the data for $E = M_Z c^2 \simeq 91.2$ GeV [51] from Z decays.

We see that $\alpha_s(E)$ does indeed decrease with E according to the formula (11.12). (For lower energies, where $\alpha_s(E)$ becomes relatively large, higher powers of $\alpha_s(E)$ in the solution of (11.11) have to be taken into account.)

In any case the variation of $\alpha_s(E)$ with E confirms the above manipulations of loop diagrams and the energy dependence generated by them. Furthermore we find that the value $\alpha_s \sim 1$ given in Chap. 6 is valid only for energies $\lesssim 1$ GeV, corresponding to the energies of quarks inside a proton or a neutron.

It is not an accident that α_s is on the order of 1 precisely for energies corresponding to the proton mass $m_p \sim 1$ GeV/c^2; we can actually turn the tables: the order of the proton mass (more precisely, about a third of the proton mass) is given by the energy/c^2 for which the energy-dependent strong fine structure constant is ~ 1! This relation is known as *dimensional transmutation*.

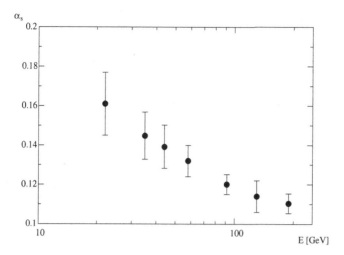

Fig. 11.5 Results of measurements of the strong fine structure constant $\alpha_s(E)$ as a function of the energy E

Exercise

11.1. The expression (11.12) for $\alpha_{s,\,measured}(E)$ can be written in the form

$$\alpha_{s,\,measured}(E) = \frac{1}{b_s \, \ln(\Lambda_{QCD}^2/E^2)}. \qquad (11.16)$$

Derive the dependence of Λ_{QCD} on α_s and Λ in (11.12), as well as on $\alpha_{s,\,measured}(E)$ and E. (Instead of the energy dependent fine structure constant, Λ_{QCD} can be considered as the fundamental parameter of the strong interaction. For $E = \Lambda_{QCD}$ $\alpha_{s,\,measured}(E)$ seems to diverge; for $\alpha_{s,\,measured} \gtrsim 1$, (11.12) is no longer valid, however.)

Derive a formula for $\alpha_{s,\,measured}(E_1)$ as a function of $\alpha_{s,\,measured}(E_2)$, E_1, and E_2. In the interval 22 GeV $< E <$ 91 GeV we have $b_s = -23/12\pi$. Determine $\alpha_{s,\,measured}(22 \text{ GeV})$ from $\alpha_{s,\,measured}(M_Z c^2) \simeq 0.12$ with the help of this formula, and compare the result to Fig. 11.5.

Chapter 12
Beyond the Standard Model

This chapter focuses on speculative extensions of the Standard Model. Determinations of the three fundamental coupling constants characterizing the three fundamental interactions indicate that, under certain assumptions, their numerical values could be identical. This motivates the idea of a Grand Unification of all three interactions, the consequences of which, such as the possible decay of protons, are sketched. Another speculative extension of the Standard Model is supersymmetry. Supersymmetry predicts the existence of many new elementary particles, which could be discovered at present or future particle physics experiments. Finally, the application of the concepts of quantum field theory to the fourth fundamental interaction, gravity, leads to internal inconsistencies. These inconsistencies can be resolved by the introduction of string theory. String theory, in turn, predicts the existence of additional spatial dimensions. We explain how the assumption of so-called compact extra dimensions can make their existence coherent with our observation of a three-dimensional space. A very speculative idea is related to the huge number of solutions of the equations of motion in string theory, which could lead to the existence of a huge number of different Universes with different properties, the so-called Multiverse.

12.1 Grand Unification

On the right-hand sides of (11.9), (11.12), and (11.13) we find the fundamental coupling α (or α_s, α_w), as well as the logarithm of the parameter Λ. Λ corresponds to an energy scale where the theory (i.e., the Feynman rules corresponding to the known particles and couplings) is possibly modified, such that integrals over energies Q^2 of virtual particles as in (11.1) should be bounded by $Q^2 \leq \Lambda^2$ (see (11.3)).

First, the fundamental couplings and Λ are unknowns in principle. After some calculation, the three equations (11.9), (11.12), and (11.13) can be brought into the form

$$\alpha = \frac{\alpha_{\text{measured}}}{1 - b\,\alpha_{\text{measured}}\,\ln\left(\Lambda^2/E^2\right)}. \tag{12.1}$$

U. Ellwanger, *From the Universe to the Elementary Particles*,
Undergraduate Lecture Notes in Physics, DOI: 10.1007/978-3-642-24375-2_12,
© Springer-Verlag Berlin Heidelberg 2012

α_{measured} and E are quantities determined by measurements, and the energy dependences of α_{measured} and $\ln\left(\Lambda^2/E^2\right)$ on the right-hand side of (12.1) cancel. Assuming a certain value for Λ, we can compute the fundamental couplings α from (12.1), since the constants b are calculable. Now we will perform this task.

The most precise measurements of $\alpha_{\text{s,measured}}$ and $\alpha_{\text{w,measured}}$ have been performed at $E = M_Z c^2$ with the results

$$\alpha_{\text{s,measured}} \simeq 0.12 \, , \qquad \alpha_{\text{w,measured}} \simeq 0.034. \qquad (12.2)$$

Substituting the values from (11.14) for b_s and b_w, and those from (12.2) for $\alpha_{\text{s,measured}}$ and $\alpha_{\text{w,measured}}$ in the corresponding equations (12.1), we find

$$\alpha_s = \alpha_w \simeq 2.14 \times 10^{-2}, \qquad (12.3)$$

if the parameter Λ satisfies

$$\ln\left(\frac{\Lambda^2}{M_Z^2 c^4}\right) \simeq 69. \qquad (12.4)$$

Hence the "fundamental" coupling constants α_s and α_w would be identical if Λ were very large, according to (12.4) for

$$\Lambda \sim 10^{17} \, \text{GeV}. \qquad (12.5)$$

This idea is called the unification of couplings. However, in a true (grand) unified theory one would expect the unification of all couplings, including that of electromagnetism.

In the framework of a Grand Unified Theory (GUT), the treatment of the electromagnetic coupling is somewhat complicated. We mentioned in Chap. 9 that the correct theory of the electromagnetic and the weak interactions is based on a SU(2) gauge symmetry and a U(1)$_Y$ gauge symmetry, where the U(1)$_Y$ gauge symmetry corresponds to a coupling constant g_1, i.e., a fine structure constant $\alpha_1 = g_1^2 \hbar/4\pi$ related to the electromagnetic coupling as in (9.35).

In a Grand Unified Theory we expect the unification of the fundamental couplings α_s, α_w, and a third coupling α_1' related to α_1 as follows:

$$\alpha_1' = \frac{5}{3}\alpha_1. \qquad (12.6)$$

The value of $\alpha_{1,\text{measured}}'$ (for $E = M_Z c^2$) is $\alpha_{1,\text{measured}}' \simeq 0.017$. The fundamental coupling α_1' depends on $\alpha_{1,\text{measured}}'$ in the same way as the couplings α_s and α_w via (12.1); it suffices to replace the parameter b by $b_1 = 41/(40\pi)$. Assuming that Λ is given by (12.5), we find for the fundamental coupling α_1'

$$\alpha_1' \simeq 2.75 \times 10^{-2}. \qquad (12.7)$$

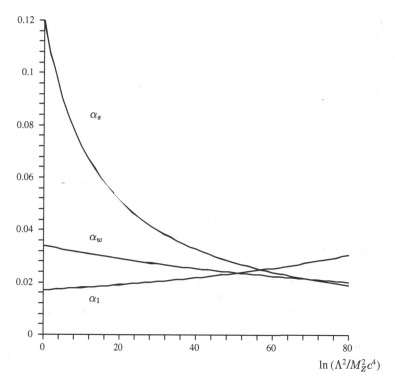

Fig. 12.1 Dependence of the three fundamental coupling constants on Λ

This is indeed close to the value in (12.3) for α_s and α_w, but not identical.

If we consider various values for Λ, it is helpful to represent the three formulas

$$\alpha_i = \frac{\alpha_{i,\,\text{measured}}}{1 - b_i\,\alpha_{i,\,\text{measured}}\,\ln\left(\Lambda^2/M_Z^2 c^4\right)} \tag{12.8}$$

in a common plot for α_i as a function of $\ln\left(\Lambda^2/M_Z^2 c^4\right)$ (with $i = 1$, s, and w; in the following we omit the prime $'$ of α_1', and use the values at $M_Z c^2$ for $\alpha_{i,\text{measured}}$). In the case of a Grand Unification, the three curves should intersect in a single point at the corresponding value of Λ. This plot is given in Fig. 12.1. We see that the three points where the three curves intersect are close to each other but do not coincide.

Finally we should add a remark on the origin of the factor 5/3 in (12.6). In fact, a Grand Unified Theory describes more than a unification of the numerical values of the coupling constants; the quarks and leptons (of a given family) are also "unified" as follows.

An important property of the strong interaction is the three colors of the quarks, which can be represented by a three-component vector. These three color components get mixed by SU(3) transformations U_{ij} as in (9.13), which is a symmetry of the strong interaction. The corresponding quantity of the weak interaction is the weak isospin,

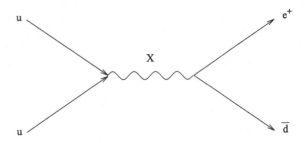

Fig. 12.2 Annihilation of two u quarks into a d̄ antiquark and a positron via an X boson in a Grand Unified Theory; this process can trigger a decay of a proton into a neutral pion and a positron

which is represented by a two-component vector. These components get mixed by the SU(2) transformations of the weak interaction. In a Grand Unified Theory, part of the quarks and leptons are combined in a $3 + 2 = 5$ component vector, the other part in an antisymmetric 5×5 matrix. The corresponding components get mixed by SU(5) transformations, which would be a symmetry of such a theory; the factor 5/3 is associated with this reasoning [52]. (There exist different theories of Grand Unification, however, which are based on different symmetry groups.)

An important consequence of this unification of quarks and leptons is that their electric charges are necessarily related: for the charges of the quarks we find indeed $+(2/3)e$ and $-(1/3)e$, where e is the positron charge. This can be considered as a strong argument in favor of such theories.

A theory of Grand Unification also contains new interactions beyond the three known interactions, resulting from the exchange of additional gauge bosons in a SU(5) gauge theory denoted as X bosons. The process sketched in Fig. 12.2 is particularly interesting: two u quarks can annihilate, and transform into a d̄ antiquark and a positron (or a μ^+). If this process takes place inside a proton, the proton decays into a neutral pion and a positron. This process has not yet been observed. This does not imply that this process never happens but, at least, that it is extremely rare. We recall the reason why the weak interaction is "weak", i.e., relatively rare: as explained in Chap. 7, the origin is the (large) mass of the $W^\pm$ bosons. The fact that the process sketched in Fig. 12.2 is so seldom that it has not yet been observed in proton decay experiments implies that the X bosons must be extremely heavy: $M_X \gtrsim 10^{16}\,\mathrm{GeV}/c^2$.

In a Grand Unified Theory, where the fundamental coupling constants are identical for a given value of Λ, the masses of the X bosons are of the order Λ/c^2. Correspondingly the absence of an observed proton decay implies

$$\Lambda \gtrsim 10^{16}\,\mathrm{GeV}, \tag{12.9}$$

which is compatible with (12.5).

In conclusion, we see that the Grand Unified Theory is very promising: it allows for the unification of the three interactions in a single comprehensive interaction (based, e.g., on a SU(5) gauge symmetry, which must be spontaneously broken into the $U(1)_Y$, SU(2), and SU(3) subgroups by an additional Higgs field such that the X bosons become massive), an explanation of the electric charges of the quarks, and

Fig. 12.3 Loop diagrams
contributing to the constant
C in (12.12)

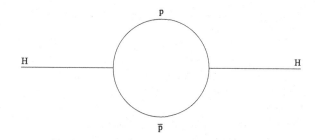

Fig. 12.3 Loop diagrams
contributing to the constant
C in (12.12)

an explanation of the ratios of the three measured coupling constants: these would follow from three equations of the form (11.9) with identical values for α and Λ but different values of the parameters b_i. The results of the calculations of the parameters b_i, which imply "nearly" a unification of the three coupling constants as shown in Fig. 12.1, seem to indicate that we are on the right track. However, the simultaneous unification of all three coupling constants is not perfect; we will return to this issue at the end of the next section.

12.2 The Hierarchy Problem and Supersymmetry

We have seen in Chap. 11 that we have to distinguish "fundamental" from measured coupling constants. The reason lies in Feynman diagrams with several vertices and loops that contribute to the processes that are used to measure these quantities. These contributions depend on the parameter Λ acting as an upper cutoff of the integrals over the energies Q of the virtual particles in the loops of the diagrams in Figs. 11.2 and 11.3.

In fact, these considerations are applicable to all parameters of the theory, including the expression (7.16) for the potential energy as a function of the Higgs field. This expression depends on two parameters μ^2 and λ_H^2. A "measurement" of these parameters corresponds to a determination of the value of the Higgs field at the minimum of the potential energy:

$$H = \mu_{\text{measured}}/\lambda_{\text{H,measured}}. \tag{12.10}$$

We know from Chap. 7 that this value amounts to $H \simeq 248\,\text{GeV}$. What type of relations exist between the parameters μ_{measured}, $\lambda_{\text{H,measured}}$ and the corresponding fundamental parameters μ, λ_H? For λ_H (a dimensionless parameter like the fine structure constants) we find a relation analogous to (12.8). In the case of μ, a parameter with units of energy, this relation is very different, however. The reason is that the corresponding diagrams (see Fig. 12.3) lead to integrals of the form

$$\int_0^{\Lambda^2} \mathrm{d}Q^2 = \Lambda^2 \tag{12.11}$$

instead of (11.3). Accordingly we find, instead of a relation of the form (11.7),

$$\mu^2_{\text{measured}} = \mu^2 - C\Lambda^2, \qquad\qquad (12.12)$$

where the constant C is calculable and on the order of 1.

This poses a big problem if Λ is on the order of 10^{16}–10^{17} GeV as in a Grand Unified Theory. Let us compare the orders of magnitude of the three terms in (12.12), starting with μ_{measured}. For $\lambda_{\text{H,measured}}$ we can assume $\lambda_{\text{H,measured}} \sim 1$; a value much larger than 1 would in fact be impossible in quantum field theory. Then (12.10) implies $\mu_{\text{measured}} \sim 250$ GeV. Now, if $\Lambda \sim 10^{16}$ GeV holds for Λ in (12.12), the parameter μ must also be on the order of $\sim 10^{16}$ GeV on the one hand, but extremely fine tuned (to a precision of 14 places!) such that the difference $\mu^2 - C\Lambda^2$ is much smaller than 10^{32} GeV^2.

However, μ is the "fundamental" parameter, and the difference between μ^2 and μ^2_{measured} (the value $-C\Lambda^2$) originates as before from quantum effects, i.e., from Feynman diagrams as in Fig. 12.3 with loops of virtual particles.

The paradox is the following: how can the fundamental parameter μ "foresee" that it has to compensate a value $-C\Lambda^2$ almost, but not completely, in (12.12)? We may assume without any difficulty that μ^2 is on the order of $(10^{16}$ GeV$)^2$ as well, but then we would obtain typically the same order for the difference $\mu^2 - C\Lambda^2$. (In fact μ knows nothing about quantum effects, i.e., about the precise value of the constant C.) We do not know any mechanism that could fix the fundamental parameter μ in a natural way such that we obtain $\mu^2 - C\Lambda^2 \ll (10^{16}$ GeV$)^2$. This problem is known as the hierarchy problem.

This problem would be solved if the constant C in (12.12) vanished. The calculation of this constant involves a sum over Feynman diagrams of the form in Fig. 12.3.

All possible particles p and antiparticles $\bar{\text{p}}$ with couplings to the Higgs boson (i.e., all massive particles) can circulate in the loop, and we have to sum over all these contributions. Hence the total contribution depends on the number and the properties of all existing elementary particles, see the tables in Chap. 10. An important observation is that the contributions of fermions with spin $\hbar/2$ to the constant C are of opposite sign to the contributions of bosons.

Now we can make the following assumption: there exist additional elementary particles, whose properties are related to those of the known particles by a new symmetry denoted as *supersymmetry* [53]: supersymmetry predicts as many new particles as the particles we know already, and that their electric charges, strong and weak interactions, and couplings to the Higgs boson are the same, but that their spin differs by $\hbar/2$. In a supersymmetric extension of the Standard Model we would find an additional boson with spin 0 for every quark and lepton (which have already been given the names squarks and sleptons), and an additional fermion with spin $\hbar/2$ (photino, gluino, gauginos, and Higgsinos) for every boson in Table 10.2 in Chap. 10.

In a supersymmetric extension of the Standard Model, the problematic equation (12.12) would be dramatically modified: now the contributions of bosons and fermions to the sum over all particles in the loop in Fig. 12.2 and hence to the constant C cancel (nearly exactly, see below)—a supersymmetric extension of the Standard

Table 12.1 Particles of the Standard Model, and the additional particles in its supersymmetric extension

Standard model	Supersymm. extension
Quarks (spin $\hbar/2$)	Squarks (spin 0)
Leptons (spin $\hbar/2$)	Sleptons (spin 0)
Photon (spin $\hbar$)	Photino (spin $\hbar/2$)
Gluon (spin $\hbar$)	Gluino (spin $\hbar/2$)
$W^{\pm}$, Z (spin $\hbar$)	Gauginos (spin $\hbar/2$)
Higgs boson (spin 0)	Higgsinos (spin $\hbar/2$)
	Additional Higgs bosons (spin 0)

Model solves the hierarchy problem. However, then the new elementary particles listed in Table 12.1 should exist. (In addition to the "partner particles" whose spin differs by $\hbar/2$ from the known particles of the Standard Model, a supersymmetric extension of the Standard Model contains additional Higgs bosons; at least three neutral ones and a charged one.)

Initially, supersymmetry predicts that the masses of the new "partner particles" should be the same as those of the known particles of the Standard Model, since their couplings to the Higgs boson are the same. This cannot be true since they have not (yet?) been discovered. This does not imply that they do not exist but, at least, that they are so heavy that their production at the most energetic of today's accelerators has not yet been possible. In fact we can understand why the masses of the new "partner particles" are larger than those of the known particles of the Standard Model if supersymmetry is spontaneously broken, similar to the SU(2) symmetry of the weak interaction. (In fact, spontaneous breaking of the SU(2) symmetry by a constant Higgs field also implies masses, amongst others of the $W^{\pm}$ and Z bosons.) The most elegant way to break supersymmetry spontaneously requires the framework of so-called *supergravity theories* [54]: in these theories, a fermionic "partner particle" exists also for the graviton, the so-called *gravitino* with spin $(3/2)\hbar$.

However, the necessary assumption of supersymmetry breaking does not allow one to make precise predictions of the masses of the new "partner particles", only that they are of about the same order; this value is denoted as M_{Susy}. (In contrast to the masses of the particles of the Standard Model, this mass is not generated by a coupling to the Higgs field.) M_{Susy} must be larger than about $100\,\text{GeV}/c^2$, since these particles have not yet been discovered.

Then we find that the contributions of bosons and fermions to the constant C originating from the diagrams in Fig. 12.2 no longer cancel exactly, but that a remainder proportional to M_{Susy}^2 is left over; in a supersymmetric theory we have to replace (12.12) by

$$\mu^2_{\text{measured}} = \mu^2 - C' M_{\text{Susy}}^2, \qquad (12.13)$$

where C' is another constant of order 1. Now the hierarchy problem—the requirement that μ has to cancel a quantum contribution much larger than μ_{measured}—is still solved if M_{Susy} is not much larger than μ_{measured}.

$M_{\text{Susy}} \sim 250\,\text{GeV}/c^2$ would imply, however, that at least some of the new particles should be discovered soon, at least in a few years, when we can expect results from the Large Hadron Collider (LHC).

Interestingly enough, a supersymmetric extension of the Standard Model also predicts that the mass of the Higgs boson (i.e., the mass of the lightest of all neutral Higgs bosons) cannot be very large: we mentioned in Sect. 7.3 that the mass of the Higgs boson cannot be predicted, since it depends on the unknown parameter λ_H in (7.16) for the potential energy $E_{\text{pot}}(H)$. In the case of a supersymmetric extension of the Standard Model, this parameter is related to the known electromagnetic and weak coupling constants, which allows an upper bound on the mass M_h of the lightest of all the neutral Higgs bosons to be computed [55, 56]:

$$M_h^2 \lesssim M_Z^2 + \frac{3m_{\text{top}}^4}{2\pi^2(248\,\text{GeV}/c^2)^2} \log\left(\frac{m_{\text{top}}^2 + M_{\text{Susy}}^2}{m_{\text{top}}^2}\right) + \ldots, \qquad (12.14)$$

where m_{top} is the top quark mass. The first term M_Z^2 in (12.14) stems from the fact that, in a supersymmetric theory, the coupling of the Z boson to the Higgs field would be nearly the same as the Higgs self-coupling, and hence the Z boson mass would be nearly the same as the lightest neutral Higgs boson mass.

In a theory with unbroken supersymmetry ($M_{\text{Susy}} = 0$), the second term in (12.14) would vanish because $\log(1) = 0$. In the realistic case of broken supersymmetry ($M_{\text{Susy}} \neq 0$) this term is due to the fact that the effects of particles of the Standard Model (such as quarks) and the new "partner particles" (such as squarks) in the diagrams in Fig. 12.2—which contribute also to M_h^2—no longer cancel exactly. The remaining effect is proportional to the fourth power of the couplings of these particles to the Higgs boson, and this coupling is proportional to the mass of these particles (see (7.19)). Hence the numerically most important contribution is due to the heaviest particle of the Standard Model, the top quark, and its partner particle, the top squark. The remaining effects of lighter particles of the Standard Model, as well as of more complicated diagrams, are indicated by dots in (12.14).

With the help of the known Z boson and top quark masses, assuming $M_{\text{Susy}} \lesssim 1\,\text{TeV}/c^2$ and taking the contributions indicated by dots into account, (12.14) implies $M_h \lesssim 130\,\text{GeV}/c^2$. This value is larger than the present experimental lower bound of $M_h \gtrsim 114\,\text{GeV}/c^2$ (from the non-discovery of a Higgs boson at LEP, see Chap. 8), but can be verified at the LHC.

(We should add that there exist theoretically more complicated supersymmetric extensions of the Standard Model with more Higgs bosons, within which the lightest Higgs boson can still be somewhat heavier and/or possess reduced couplings, which would complicate its detection.)

If the particles predicted within a supersymmetric extension of the Standard Model exist, we have to compute anew the parameters b_i in (12.8) of the previous section: the new particles would also circulate in the loops in the diagrams in Figs. 11.2 and 11.3, which modifies the numerical values of the parameters b_i. Instead of the values in (11.14) and $b_1 = 41/(40\pi)$ we would obtain

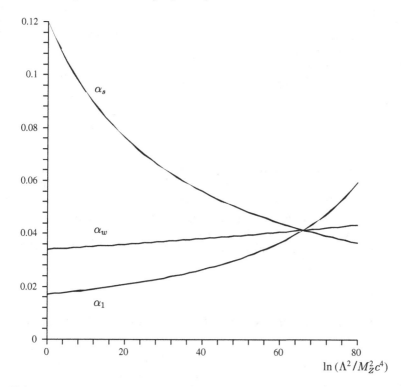

Fig. 12.4 Λ dependence of the three fundamental couplings in a supersymmetric extension of the Standard Model

$$b_s = -\frac{3}{4\pi}, \qquad b_w = \frac{1}{4\pi}, \qquad b_1 = \frac{33}{20\pi}. \qquad (12.15)$$

Instead of the curves in Fig. 12.1 we would find (for the same values of the measured α_i) the situation in Fig. 12.4.

Now the three curves intersect in a single point! This means that the assumption of Grand Unification is possible if it is combined with supersymmetry. The value of Λ for which the three fundamental couplings would be identical according to Fig. 12.4 is

$$\Lambda \sim 2 \times 10^{16} \, \text{GeV}. \qquad (12.16)$$

This value barely satisfies the inequality (12.9) following from the absence of an observed proton decay.

It is hard to believe that the result of Fig. 12.4 is a mere accident. Many particle physicists consider the common intersection point of the curves as a strong argument in favor of the validity of the two hypotheses Grand Unification and supersymmetry.

Finally we find among the particles predicted within a supersymmetric extension of the Standard Model on the right-hand side of the Table 12.1 in most cases a

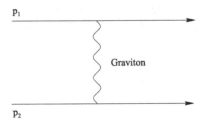

Fig. 12.5 Exchange of a graviton between two particles generating the usual attractive gravitational force

neutral stable particle with mass $\sim M_{\text{Susy}}$ (the lightest among the photino, the neutral gauginos, and Higgsinos). Among the open questions in cosmology is the nature of dark matter. One possibility is indeed that it consists of elementary particles with precisely these properties: neutral, stable, and relatively massive. Accordingly, supersymmetry could also solve the problem of the nature of dark matter.

In a few years we will know more about whether this theory actually does describe nature.

12.3 Quantum Gravity, String Theory, and Extra Dimensions

In principle we can apply the formalism of quantum field theory to gravity as well, and describe the gravitational force (or gravitational interaction) between two objects or two particles p_1 and p_2 by the exchange of a graviton with spin $2\hbar$ as in Fig. 12.5.

The interaction or rather amplitude $A(\theta)$ generated by this diagram corresponds, up to a constant, to the one in (quantum) electrodynamics discussed in Chap. 5. (We have to replace the electromagnetic fine structure constant by the gravitational coupling.) It follows that the gravitational force generated by this diagram depends on the distance r between the objects precisely like the electric force (up to a constant), $\left|\vec{F}_{\text{Grav}}(r)\right| \sim 1/r^2$—again the simplest Feynman diagram (without loops) reproduces the result of classical physics, here the gravitational interaction originating from the metric (3.34).

In quantum gravity there exist vertices where a particle couples to two (or more) gravitons. (Such vertices do not exist with electrons and photons.) Accordingly, diagrams as in Fig. 12.6 contribute to the interaction between two particles in quantum gravity.

The calculation of the diagram in Fig. 12.6 requires integration over the energies Q of the gravitons in the loop, and again the integral has to be cut off at the upper boundary by Λ^2. If we denote the total energy of the particles p_1 and p_2 by E, the contribution of this diagram to the amplitude is on the order of

$$\kappa^2 \frac{E^2}{\hbar c^3} \ln\left(\Lambda^2/E^2\right) A^{(2)}(\theta). \qquad (12.17)$$

Fig. 12.6 Graviton loop diagram giving a contribution to the amplitude of the form (12.17)

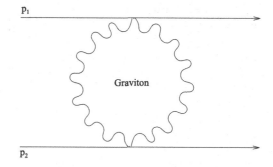

p_1

Graviton

p_2

Again $A^{(2)}$ describes the dependence on the scattering angle θ. This contribution seems to be proportional to $1/\hbar$ instead of $\hbar$; however, two powers of $\hbar$ are "hidden" in E^2, since E is proportional to $\hbar$, see (4.8). Hence we obtain for the sum of the contributions of the diagrams in Figs. 12.5 and 12.6

$$A(\theta) = \kappa A^{(1)}(\theta) + \kappa^2 \frac{E^2}{\hbar c^3} \ln\left(\Lambda^2/E^2\right) A^{(2)}(\theta), \qquad (12.18)$$

where κ denotes the gravitational constant defined in (2.5). Here κ replaces the fine structure constant α in the analogous equation (11.4), the sum of the diagrams in Figs. 11.1 and 11.2 of the electromagnetic interaction.

Equation (12.18) differs from (11.4) in that

(a) the θ dependences of $A^{(1)}(\theta)$ and $A^{(2)}(\theta)$ are no longer the same, and
(b) the energy dependence of the contribution of the loop diagram in Fig. 12.6 differs from that in the diagram in Fig. 12.5.

However, the contribution of the loop diagram in Fig. 12.6 is numerically relevant only for energies $E \sim \sqrt{\hbar c^3/\kappa} \sim 2.4 \times 10^{18}$ GeV, for which the second term in (12.18) becomes of the same order as the first term. In fact, at such energies all multiloop diagrams become of the same order. Point-like particles—only for these does the diagram in Fig. 12.6 lead to a contribution of the form (12.17)—with such energies will not be able to be produced in the foreseeable future. In principle, even a single particle could possess such an energy at rest (according to $E = mc^2$) if it were extremely massive: its mass would have to have a value of

$$M_{\text{Planck}} = \sqrt{\frac{\hbar}{c\kappa}} \simeq 2.4 \times 10^{18} \text{ GeV}/c^2, \qquad (12.19)$$

which is known as the Planck mass. (Occasionally the value in (12.19) multiplied by $\sqrt{8\pi}$ is used.) However, it amounts to more than 10^{16} times the mass of the heaviest known particle, the top quark with a mass of about ~ 173 GeV$/c^2$.

Whereas it is therefore practically impossible to verify experimentally the contributions of loop diagrams to the gravitational interaction in quantum gravity, it is no longer possible, for reasons (a) and (b) above, to compensate the Λ dependence of

Fig. 12.7 Open and closed
strings of length l

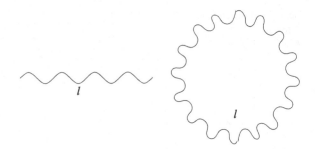

the contribution (12.17) by the introduction of a "measured" gravitational constant
κ_{measured}; quantum gravity is not renormalizable. The limit $\Lambda \to \infty$—keeping the
measured gravitational constant κ_{measured} fixed—is no longer possible, and measurable quantities depend—in contrast to a renormalizable theory—in principle on Λ:
the naive combination of gravity (in the form of the well confirmed theory of general
relativity) and quantum field theory (well confirmed in particle physics) leads to
inherent contradictions.

Therefore we need a justification for the introduction of a finite parameter Λ. Such
a justification corresponds to a modification of the Feynman rules, i.e., a modification
of quantum gravity. Without such a modification, quantum gravity is not a reasonable
theory.

Researchers have worked on possible modifications of these fundamental theories,
and the most promising consists in replacing quantum gravity and, simultaneously,
the complete Standard Model of particle physics, by *string theory*.

In string theory, each elementary particle—fermions, bosons, the graviton, etc.—
is replaced by a string of length l. As sketched in Fig. 12.7, these strings can be open
or closed.

This does not contradict the fact that no finite size (or inner structure) of quarks
or leptons has been detected: we only know that a possible inner structure must be
smaller than about 10^{-18} m; hence all we can conclude is that the length l of the
strings must also be smaller than about 10^{-18} m.

In string theory (in its simplest version) the length l can be determined: replacing
the graviton by a string, its exchange generates an interaction of the known form of
gravity, with a gravitational constant κ given by

$$\kappa = l^2 c/\hbar. \tag{12.20}$$

Since we know the value of κ, we obtain from the known values of c and $\hbar$

$$l \simeq 8 \times 10^{-35}\,\text{m}, \tag{12.21}$$

far below 10^{-18} m.

A string can oscillate but, given a string of finite length l, only oscillations of certain
frequencies are possible. (This allows the generation of notes of definite pitches by
string instruments.) In (relativistic) string theory, the energy of an oscillating string

depends on its frequency similar to (4.8), and accordingly a string of finite length can only be in states of definite energy. In string theory, these states correspond to different particles, whose masses m are related to the energy by the known formula $m = E/c^2$.

If we want to describe all elementary particles as well as all interactions in terms of a string theory, it should be possible to identify all particles of the Standard Model—quarks, leptons, and bosons (and possibly the additional particles predicted by supersymmetry)—with states of oscillating strings. We find, however, that the energy differences between the different oscillating states of a string are of the order Λ, where the parameter Λ is related to the string length l by

$$\Lambda = \hbar c / l. \tag{12.22}$$

The value of Λ following from (12.22) (with l from (12.20)) is equal to the energy corresponding to the Planck mass

$$\Lambda = M_{\text{Planck}} \, c^2 \simeq 2.4 \times 10^{18} \, \text{GeV}, \tag{12.23}$$

therefore the particles of the Standard Model can correspond only to the oscillating states of lowest possible energy ~ 0 (in multiples of Λ).

In principle there exist so-called bosonic string theories, where all oscillating states are bosons (with various spins, which are all integer multiples of $\hbar$), and superstring theories, containing additional fermionic oscillating states. Since the Standard Model contains fermionic particles, only superstring theories can be realistic.

In a string theory the Feynman rules are modified: here all vertices depend on the energies Q of the participating particles (strings) in the form of a decreasing exponential function $\exp(-Q^2/\Lambda^2)$. Thus all integrals appearing in loop diagrams over functions $f(Q^2)$ (like the integral in (11.1), where $f(Q^2)$ is given by $1/(Q^2 + m_e^2)$), are to be replaced by integrals of the form

$$\int_0^\infty dQ^2 e^{-Q^2/\Lambda^2} f(Q^2). \tag{12.24}$$

Owing to the strong decrease of the exponential function $\exp(-Q^2/\Lambda^2)$ for $Q^2 \gg \Lambda^2$, the results of the integrals (12.24) correspond approximately to those obtained by an upper cutoff Λ^2 (up to terms containing negative powers of Λ^2):

$$\int_0^\infty dQ^2 e^{-Q^2/\Lambda^2} f(Q^2) \simeq \int_0^{\Lambda^2} dQ^2 \, f(Q^2). \tag{12.25}$$

This is one of the main advantages of string theories: now all integrals appearing in loop diagrams are automatically finite. A parameter Λ^2 does not have to be introduced ad hoc as in (11.3); it is even calculable and useful, as in (12.17), for applications of

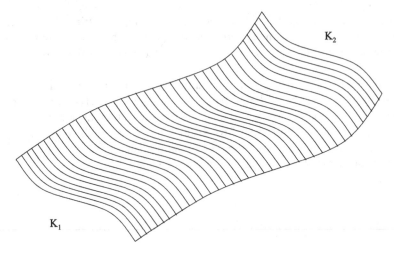

Fig. 12.8 World surface covered by an open string propagating from a configuration K_1 to a configuration K_2

quantum field theory to gravity. The problem of quantum gravity in the limit $\Lambda \to \infty$ no longer exists; in string theories the contradiction between quantum field theory and gravity is resolved.

In fact the exponential function in (12.24) modifies all vertices of interactions in the Standard Model as well. However, owing to the enormous value of Λ (and $\exp(-\varepsilon) \to 1$ for $\varepsilon \to 0$) this does not lead to observable consequences, since we can carry out experiments only at energies $Q \ll \Lambda$.

On the other hand, in string theory the value (12.23) for Λ would also have to be used in Sects. 12.1 and 12.2, in particular in (12.16) following from the Grand Unification of coupling constants in a supersymmetric extension of the Standard Model.

On a logarithmic scale, the two values of about 10^{18} GeV and about 10^{16} GeV for Λ would not lie far apart; the relative proximity of these values seems to indicate that we are on the right track. However, a plausible reason for the remaining relative factor 100 remains to be found.

It is not generally true, however, that we always obtain a consistent theory in string theory. The problem appears already in the calculation of the simplest possible process: let us assume that, at a time $t = 0$, a string is in a configuration K_1. Now we want to compute the probability of finding the string in a configuration K_2 at a given later time. According to the rules of quantum mechanics—valid for string theory as well—we have to sum over all possible ways of reaching a configuration K_2 from K_1. For point particles this is a relatively simple task; however, in string theory this task is considerably more complicated, as becomes apparent from Fig. 12.8 (for, e.g., open strings).

Between the configurations K_1 and K_2, the string covers a so-called world surface. Now we have to sum over all world surfaces bounded by K_1 and K_2. Summing over

Fig. 12.9 A cylinder, symmetric around the axis, on a partially compactified two-dimensional surface

world surfaces corresponds to an integration over all possible curvatures of world surfaces bounded by K_1 and K_2. Curvatures of world surfaces are described as in Sect. 3.2 by *metrics* g_{ij}, which are 2×2 matrices here, since world surfaces are two-dimensional spaces. (This metric is not to be confused with the metric of the space–time in which the string propagates!) These metrics can be characterized by parameters, and we have to integrate over all such parameters.

Now we observe, however, that these integrals are usually infinite. In fact, owing to the so-called conformal anomaly, the coefficients of the infinite contributions depend on the dimension of space–time in which the string propagates: in the case of bosonic strings the coefficient of the infinite contributions is proportional to $d - 26$, and in the case of the more interesting superstrings, proportional to $d - 10$. Hence the absence of the infinite contributions requires (for the more interesting superstrings) that the dimension d of space–time is $d = 10$!

At first sight this contradicts the fact that (according to the theory of special relativity) we live in a $d = 3 + 1 = 4$-dimensional space–time. However, we can admit additional dimensions under the condition that space is "compact" in these additional directions. In geometry, "compact" means that the extension of space is finite in the corresponding direction. To better understand this concept it is helpful, as in Chap. 3, to imagine a two-dimensional space. The surface of a sphere is an example of a two-dimensional space that is compact in all directions: the total surface is finite (in contrast to the surface of a plane), and straight motions in any arbitrary direction lead back to the point of departure; infinite distances do not exist on this surface.

There exist two-dimensional spaces which are partially compact: imagine a sheet of paper (representing, initially, an infinite flat plane) and roll it up to form a tube with diameter D. The length of the tube is still infinite, but it is of finite circumference $U = \pi D$: this space is not compact along the tube, but compact around the tube.

Even if the surface of this tube is a two-dimensional space, it is difficult to distinguish it from a line (a one-dimensional space) if its diameter D is very small (i.e., if the tube resembles a straw) and observed from a very large distance.

In fact a compact dimension, corresponding to a direction in which space has only a finite extension U, can be detected only if we can resolve structures smaller than U. Let us compare the possible motions of two different objects on the surface of a straw: an ant, much smaller than U, can move in two different directions: along the straw or around the straw—it can recognize both dimensions. On the other hand a cylinder, symmetric around the axis (see Fig. 12.9), would correspond to an object whose size along the compact dimension is equal to the extension U of the dimension. The cylinder can move, like the ant, along the straw. However, after a motion around the axis of the straw it remains unchanged; the black surface in Fig. 12.9 remains the same. Accordingly it cannot "experience" the compact dimension; the only possible

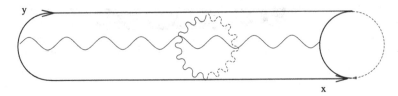

Fig. 12.10 Wave solutions of the Eq. 12.26 on a partially compactified surface

motion inside the two-dimensional space that changes its state is a one-dimensional motion along the axis. For this reason the cylinder perceives only a one-dimensional world. In fact it can rotate around the axis, but the rotational velocity and the resulting internal energy would be invariable properties of the cylinder. Generally speaking, compact dimensions are "invisible" for objects large compared to the extension of space in this direction (i.e., which fill this direction completely).

We mentioned at the end of Sect. 4.2 that we can resolve spatial structures of size Δ only if we carry out experiments at an energy $E \gtrsim \hbar c/\Delta$. This also holds for the detection of compact dimensions, in which a space possesses only a finite extension U. How a compact dimension can be detected with the help of sufficiently large energies follows from field theory and the Klein–Gordon equation (4.1). In our example corresponding to Fig. 12.9, space has two dimensions, where the direction along the tube can be denoted by x and the direction around the tube by y. Correspondingly we will consider the (for simplicity massless) Klein–Gordon equation (4.1) for fields $\Phi(\vec{r}, t)$ depending only on x, y, and t:

$$\left(\frac{\partial^2}{\partial t^2} - c^2 \left(\frac{\partial^2}{\partial x^2} + \frac{\partial^2}{\partial y^2}\right)\right)\Phi(x, y, t) = 0. \tag{12.26}$$

$\Phi(x, y, t)$ can depend on x as well as on y in the form of waves as sketched in Fig. 12.10.

Now it is important that the wavelengths around the tube must satisfy

$$\lambda = \frac{U}{n}, \quad n = \text{integer}, \tag{12.27}$$

such that after each turn around the tube a wave crest encounters a wave crest, and a wave trough a wave trough (see the corresponding condition (5.42) in the Bohr atomic model). For this reason, the solutions of the equation (12.26) for $\Phi(x, y, t)$ are of the form

$$\Phi(x, y, t) = \Phi_0 \cos(\omega t - kx) \cos(2\pi n y/U), \tag{12.28}$$

since, after a full turn $y \to y + U$ and $\cos(2\pi n y/U + 2\pi n) = \cos(2\pi n y/U)$, the last factor is the same. Replacing this expression for $\Phi(x, y, t)$ in (12.26) we find that ω has to satisfy

$$\omega^2 = k^2 c^2 + \left(\frac{2\pi nc}{U}\right)^2.$$ (12.29)

This corresponds to the relation $\omega^2 = k^2 c^2 + m_n^2 c^4/\hbar^2$ for massive particles (see below (4.11)) with masses given by

$$m_n = \frac{2\pi\hbar n}{Uc}!$$ (12.30)

Since n can assume all possible positive integer values, an entire "tower" of particles with masses m_n, $n = 0, \ldots, \infty$, exists in this world. These particles are called Kaluza–Klein states [57, 58]. The lightest among these particles with $m_0 = 0$, corresponding to $n = 0$, would also exist in a world without the extra dimension y. The presence of the infinite number of additional particles with masses $m_n = 2\pi n\hbar/(cU)$—with the same charges as the particle with $m_0 = 0$—indicates the presence of the extra compact dimension y.

The lightest additional particle (corresponding to $n = 1$) can be produced only if we carry out experiments at an energy of at least $E = m_1 c^2 = 2\pi \hbar c/U$, which, according to (4.12), corresponds to a resolution power $\Delta \sim U/(2\pi)$ (on the order of the "radius" of the extra dimension).

These considerations remain valid for higher-dimensional spaces where some of the dimensions are compact: it is in fact possible that our space–time is ten dimensional, if six of the ten dimensions are compact and of a microscopic extension smaller than about 10^{-18} m: with the help of the maximum presently available energies of about 1000 GeV we could have detected Kaluza–Klein states of the known particles—and hence the presence of extra dimensions—only if their radius were larger than about 10^{-18} m.

In a ten-dimensional string theory it is natural (but not obligatory) that the circumference U of the six compact dimensions is on the order of the string length l, which, given its value (12.21), would be far too small to be detected via the existence of the Kaluza–Klein states. Accordingly it is quite possible that the fundamental theory is a ten-dimensional string theory.

Above we mentioned that, in a string theory, the known elementary particles of the Standard Model should be identified with the oscillatory states of lowest energy. Of course this holds for the four-dimensional theory, obtained under the assumption of so-called "compactification" of six of the ten dimensions. Now we find that the number and properties of the particles of the four-dimensional theory depend on the shape of the compactification of the six dimensions; by "shape of the compactification" we mean the curvature and additional properties (such as possible singularities, corresponding to so-called orbifolds) of the compact six-dimensional space. Many present studies are concerned with the search for all possible shapes of such compactifications, in order to find the particles and interactions of the Standard Model in the effective four-dimensional theory.

To date we know about five different ten-dimensional superstring theories. They differ in the number of open and closed strings (some contain exclusively closed strings, but theories with open strings always contain closed strings as well) and in the possible oscillations and rotations of the strings. (A rotating string, i.e., a string with angular momentum, corresponds to a particle with spin.) A common property of all string theories is the presence of a state (a closed string) with vanishing mass and spin $2\hbar$, corresponding to a graviton. Hence, all these theories include a description of gravity that includes quantum gravity, and resolve the contradiction between the theory of general relativity and quantum field theory.

After compactification of some of the ten dimensions we often find the same states in more than one of the five superstring theories; such relations are called *dualities*. In a compactified superstring theory there exist two kinds of "towers" of states:

(a) the various "ordinary" oscillatory states of the string, whose energy differences Λ are related to the string length l as in (12.22), and
(b) the Kaluza–Klein states with masses as in (12.30), which now correspond to strings "wrapped" around the compact dimension.

Often we find the same states in different compactified superstring theories, but their origins corresponding to (a) or (b) are different—such theories are called *dual* to each other. Such dualities indicate that the 5 different superstring theories are possibly nothing but different descriptions of a single (but still unknown) more fundamental theory denoted occasionally as *M theory*.

A particularly interesting form of duality relates quantum effects in open superstring theories to classical effects in closed superstring theories. To understand this connection, we first have to recall the definition of quantum effects in quantum field theory: in quantum field theory, various Feynman diagrams contribute to the computation of the probability of a given process. The Feynman diagrams without loops reproduce the results of classical physics, as we saw in Sect. 5.3 in the example of the calculation of the probability $P(\theta)$. Feynman diagrams with loops lead to additional contributions, which are, however, always proportional to higher powers of Planck's constant $\hbar$ (the power of $\hbar$ is equal to the number of loops). For this reason the contributions from loop diagrams are denoted as quantum effects; the results of classical physics, for which Planck's constant $\hbar$ plays no role, are re-obtained in the limit $\hbar \to 0$.

Now, what corresponds to "loop diagrams" of quantum field theory in string theory? "Loop diagrams" in open string theory are world surfaces with holes, and loop diagrams in closed string theory are world surfaces with "handles"! The simplest loop diagram in open string theory is the world surface with a hole shown in Fig. 12.11, which contributes to the process represented in Fig. 12.8: this diagram corresponds to a quantum effect in (open) string theory.

In string theory we re-obtain the point particles of quantum field theory if we contract all strings of length l to points, corresponding to $l = 0$. If we contract all strings in Fig. 12.11 we do indeed obtain a loop diagram for the corresponding point particles (similar to the one in Fig. 12.3). It is important that the curvature and the form of the boundary of a world surface are irrelevant for the number of holes of a world surface; in any case we have to integrate over all possible curvatures of a world surface (in the sense described above), but separately for world surfaces with

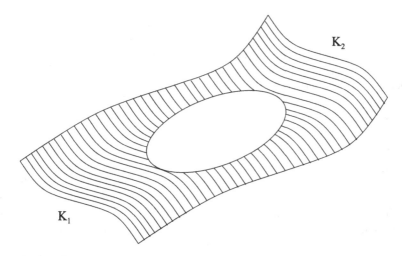

Fig. 12.11 A world surface with a hole contributing to the process in Fig. 12.8 corresponding to a loop diagram in open string theory

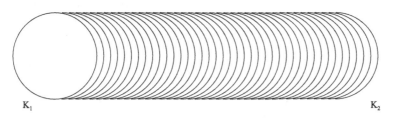

Fig. 12.12 Deformation of the world surface in Fig. 12.11, which corresponds now to the propagation of a closed string from a configuration K_1 to a configuration K_2

a given number of holes. For this reason, after a deformation of its boundaries and its curvature, the world surface in Fig. 12.11 (a quantum effect in open string theory) is equivalent to the world surface in Fig. 12.12, where, e.g., K_1 corresponds to the exterior and K_2 to the inner boundary in Fig. 12.11.

The world surface in Fig. 12.12 has a completely different interpretation in the context of *closed* string theory: the boundary K_1 corresponds to a configuration of a *closed* string at a given time, and the boundary K_2 to a closed string at a later time. Then Fig. 12.12 corresponds merely to a closed string propagating from K_1 to K_2, analogously to the process of an open string in Fig. 12.8, at the *classical* level (i.e., without quantum effects) since the world surface has no "handles". In fact, if we contract the circles in Fig. 12.12 (which correspond to closed strings) we obtain just a line—without loops! (If we contract lines directed from K_1 to K_2, corresponding to open strings, we still obtain a loop.)

The fact that one and the same world surface (i.e., integrals over the metrics) has two different interpretations in string theory is a particular kind of duality. Since one interpretation corresponds to a classical effect, but the other to a quantum effect, classical and quantum physics seem to be inseparably connected in string theory.

However, quantum effects are understood in string theory only approximatively in a series expansion in the number of holes (or handles) of world surfaces. If we neglect quantum effects—whereupon we have the theory well under control—string theory "suffers" from the fact that its equations possess an enormous number of possible solutions.

In order to understand the notion "possible solution" we have to recall that the different oscillatory states of a string correspond to ordinary particles, and hence to ordinary fields. Now it is important how the potential energy depends on all these fields. In Sect. 7.3 we introduced the potential energy as a function of the Higgs field H (see (7.2) and Fig. 7.9). The corresponding expression had two minima, which are in fact physically equivalent since they transform into each other under the symmetry transformation $H \rightarrow -H$. However, in principle every minimum of the potential energy corresponds to a stable state, denoted also as a "solution" of the theory (a constant solution of the so-called equations of motion such as the Klein–Gordon equations including coupling terms).

First, in string theory we find a large number of such fields (the precise number depends on the form of the compactification of six of the ten dimensions), and second, we find that the potential energy as a function of each of these fields possesses a large number of minima. It may even happen that the potential energy is independent of some of the fields—then, every constant value of these fields corresponds to a solution. (Such fields are denoted as *moduli*.) The number of all possible solutions is estimated to lie in the range 10^{500}–10^{1500}. The picture of the potential energy with its enormous number of minima is also denoted as a "landscape".

A few years ago, this number strongly increased again through the discovery of so-called D-branes in open string theories [59]: the notion "D-brane" is a generalization (with D an integer) of the notion "membrane", which means a two-dimensional surface. A D-brane is a D-dimensional space inside the nine-dimensional space of superstring theories (time does not count here)—similar to a two-dimensional space (a plane) that can exist inside our three-dimensional space. This D-dimensional space is singled out by the fact that some particles and fields exist only in this subspace, and not in the full nine-dimensional space. This could hold, for instance, for all particles (or fields) of the Standard Model, which would then live on a 3-brane, since our space possesses three dimensions. This possibility would render unnecessary the compactification of six of the nine spatial dimensions in superstring theories. However, in superstring theories the gravitons live always in ten space–time (i.e., nine spatial) dimensions. If the extra six dimensions are, at least partially, not compact or have a large circumference, this can imply that the dependence of the gravitational force on the distance differs from its $1/r^2$-like behavior at small distances. This behavior is verified down to distances of about 1 mm, but discoveries of deviations at smaller distances could indicate that our Universe is a 3-brane embedded in a higher-dimensional space–time.

Let us return to the enormous number of possible solutions of superstring theory, which correspond to different values of fields in different minima of the potential energy. First we recall the implication of the value of the Higgs field in the minimum of the potential energy in the Standard Model: the masses of the $W^\pm$ and Z bosons, as well as the masses of the quarks and charged leptons, are proportional to this value. A generalization of this phenomenon would take place in every minimum of the potential energy, which would correspond to a different solution (value) for a constant Higgs field: in each minimum, the particles would have different masses. In addition, coupling constants (fine structure constants, Yukawa couplings, and even the gravitational coupling) may depend on fields, and would correspondingly possess different values in different "solutions". Finally, even the cosmological constant (the value of the potential energy in the corresponding minimum) would be different in each solution—this means that the Universe, the existing particles, and interactions would differ dramatically in each different solution.

Hence, why is our known Universe—corresponding to the Standard Model of particle physics and the corresponding values of its parameters, the gravitational and cosmological constant—the one realized in nature? Such a question has been asked before in one form or another, and a possible answer is the so-called *anthropic principle* [60]: this principle states that the properties of the fundamental laws of nature have to be such that intelligent life is possible—otherwise nobody would be there to ask the question.

The necessary conditions on the parameters of the Standard Model of particle physics, including gravity, for the development of intelligent life, at least in the form known to us, is the subject of current discussions; among other things, the following considerations play a role in this context:

(a) In order that the relatively complex carbon nuclei $^{12}_{6}C$ can be generated in nuclear reactions (and remain stable) in the interior of stars, the electromagnetic and strong forces (between baryons) have to assume their actual values with a precision of about 1%.

(b) In order that massive stars with their planetary systems can be formed, the gravitational force between atoms and nuclei must be much smaller than all the other interactions.

(c) In order that enough time for the formation of stars, planets, and life is available (and the Universe neither collapses nor dilutes too much before), the value of the cosmological constant must not be too large [61].

However, the anthropic principle alone is not a complete answer to the question about the origin of the fundamental parameters; a complete answer based on the anthropic principle requires that we have the choice among many different parameters, i.e., that correspondingly many different Universes were or are realized.

On one hand, the (speculative!) idea of the "landscape" furnishes a possible model for this idea [62]: at the beginning, the fields of superstring theory could sit in different minima of the potential energy in different regions of the Universe. Since the gravitational and cosmological constants are typically very different in these different regions, the cosmological development proceeds differently. This way, $10^{500}-10^{1500}$ different Universes come into being (we also talk about a *multiverse*), most of which collapse or dilute infinitely within fractions of a second, however; we just happen to live in the one (or one of those) that, owing to its lifetime and properties, allows for intelligent life.

On the other hand, the idea of the "landscape" is based on a crude approximation of string theory, namely on the large number of possible solutions when quantum effects are neglected (or possible modifications in the framework of a still more fundamental theory). To justify the neglect of quantum effects, one could use the phenomenon of duality discussed above, according to which one observes, at least in several examples, that quantum effects in one of the five superstring theories are equivalent to "classical" effects (without world surfaces with holes, i.e., without quantum effects) in another superstring theory. In any case it is not clear at present whether superstring theories can be used to justify the anthropic principle, according to which the fundamental parameters could *not* be deduced, i.e., computed, from a fundamental theory.

Obviously it would be very interesting to better understand superstring theories— or a possibly even more fundamental theory, unifying all five superstring theories. The possibility that there is a basically unique theory, describing simultaneously all particles and interactions of the Standard Model, including gravity, is enough motivation in order to make every effort in this direction, including the development of the new mathematical methods that will probably be required.

Exercise

12.1. Compare the expression (5.10) for the electric force between two objects with charges q and q' to the expression (5.11) for the gravitational force between two objects with masses m and M: for which values of the masses $M = m$ of two particles of charge e is the modulus of the electric force equal to the modulus of the gravitational force between them? Give these masses in GeV/c^2, and compare the result to the value of the Planck mass (12.19).

Appendix A

A.1 Relevant Constants and Abbreviations

Speed of light	$c = 299792458 \, \mathrm{m \, s^{-1}}$
Newton's gravitational constant	$G \simeq 6.674 \times 10^{-11} \, \mathrm{m^3 \, kg^{-1} s^{-2}}$
$\kappa = 8\pi \, G/c^2$	$\kappa \simeq 1.866 \times 10^{-26} \, \mathrm{m \, kg^{-1}}$
Planck constant	$h \simeq 6.62607 \times 10^{-34} \, \mathrm{J \, s}$
$\hbar = h/2\pi$	$\hbar \simeq 1.054572 \times 10^{-34} \, \mathrm{J \, s}$
Charge of a positron	$e \simeq 1.602177 \times 10^{-19} \, \mathrm{C}$
Electric fine structure constant	$\alpha \simeq 7.29735 \times 10^{-3} \simeq 1/137.04$
Permittivity of the vacuum	$\varepsilon_0 \simeq 8.85419 \times 10^{-12} \, \mathrm{C \, V^{-1} m^{-1}}$
	$= 8.85419 \times 10^{-12} \, \mathrm{C^2 \, kg^{-1} m^{-3} \, s^2}$
Electron mass	$m_e \simeq 0.510999 \, \mathrm{MeV}/c^2$
Proton mass	$m_p \simeq 938.272 \, \mathrm{MeV}/c^2$
Neutron mass	$m_n \simeq 939.565 \, \mathrm{MeV}/c^2$
Lengths	$1 \, \mathrm{ly} \simeq 0.9461 \times 10^{16} \, \mathrm{m}$
	$1 \, \mathrm{pc} \simeq 3.08568 \times 10^{16} \, \mathrm{m}$
	$1 \, \mathrm{nm} = 10^{-9} \, \mathrm{m}$
	$1 \, \text{Å} = 10^{-10} \, \mathrm{m}$
	$1 \, \mathrm{fm} = 10^{-15} \, \mathrm{m}$
Energy and mass	$1 \, \mathrm{J} = 1 \, \mathrm{kg \, m^2 \, s^{-2}}$
	$1 \, \mathrm{eV} \simeq 1.602177 \times 10^{-19} \, \mathrm{J}$
	$1 \, \mathrm{eV}/c^2 \simeq 1.782662 \times 10^{-36} \, \mathrm{kg}$
Powers of Ten	$1 \, \mathrm{k} = 10^3, \; 1 \, \mathrm{M} = 10^6,$
	$1 \, \mathrm{G} = 10^9, \; 1 \, \mathrm{T} = 10^{12}.$

U. Ellwanger, *From the Universe to the Elementary Particles*,
Undergraduate Lecture Notes in Physics, DOI: 10.1007/978-3-642-24375-2,
© Springer-Verlag Berlin Heidelberg 2012

A.2 Useful Internet Addresses

Satellite Experiments for the Measurement of the Cosmic Background Radiation:
  WMAP: `map.gsfc.nasa.gov`
  Planck: `www.rssd.esa.int/index.php?project=Planck`
Experiments for the Detection of Gravitational Waves:
  GEO600: `www.geo600.org`
  LIGO: `www.ligo-la.caltech.edu`
  TAMA: `tamago.mtk.nao.ac.jp`
  Virgo: `www.virgo.infn.it`
Particle Accelerators:
  CERN (LEP, LHC): `www.cern.ch`
  Fermilab (Tevatron): `www.fnal.gov`
Overview of the Properties of Known Elementary Particles:
  Particle Data Group: `pdg.lbl.gov`

Solutions to Exercises

Chapter 1

1.1 We look for a maximum of the function $E_{binding}(A, Z)$ (neglecting $\delta(N, Z)$, and replacing $N = A - Z$). Requiring the derivative of $E_{binding}(A, Z)$ with respect to Z to vanish gives us the desired formula for $Z(A)$:

$$Z(A) = \frac{A}{2 + [a_c/(2a_a)]A^{2/3}}. \tag{A.1}$$

For $A = 238$ we obtain, using $a_c/a_a \simeq 0.030$, $Z(238) \simeq 92.4 \simeq 92$, which corresponds to the chemical element uranium.

Chapter 2

2.1 First it is helpful to rewrite (2.6), (2.7), as we did in (2.8), in terms of the function $H(t) = \dot{a}(t)/a(t)$: Equation 2.6 becomes

$$3H^2(t) = \kappa c^2 \varrho(t), \tag{A.2}$$

and (2.7) becomes

$$2\dot{H}(t) + 3H^2(t) = -\kappa p(t) = -\kappa c^2 w \varrho(t), \tag{A.3}$$

owing to the assumed relation between $p(t)$ and $\varrho(t)$. After substituting $\varrho(t)$ from (A.2) in (A.3), we find (A.3) becomes

$$\dot{H}(t) = -\frac{3}{2}(1 + w)H^2(t). \tag{A.4}$$

U. Ellwanger, *From the Universe to the Elementary Particles*,
Undergraduate Lecture Notes in Physics, DOI: 10.1007/978-3-642-24375-2,
© Springer-Verlag Berlin Heidelberg 2012

Now we have to distinguish the cases $w \neq -1$ and $w = -1$:

(a) $w \neq -1$:

The ansatz $H(t) = k/t$ (where k is a constant to be determined) leads to $k = 2/[3(1 + w)]$. From $\dot{a}(t) = H(t)\, a(t)$ we obtain

$$a(t) = a_0 t^{2/(3(1+w))}, \tag{A.5}$$

and for $\varrho(t) = [3/(\kappa c^2)]\, H^2(t)$ (from (A.2))

$$\varrho(t) = \frac{4}{3\kappa c^2 (1 + w)^2 t^2}, \tag{A.6}$$

as well as

$$p(t) = wc^2 \varrho(t) = \frac{4w}{3\kappa (1 + w)^2 t^2}. \tag{A.7}$$

(b) $w = -1$:

Equation (A.4) simplifies to $\dot{H}(t) = 0$ with the general solution $H(t) = k$. From $\dot{a}(t) = H(t)\, a(t)$ we obtain $a(t) = a_0 e^{kt}$, and for $\varrho(t)$ we find $\varrho(t) = [3/(\kappa c^2)]k^2$, and furthermore $p(t) = -(3/\kappa)k^2$. Now $\varrho(t)$ and $p(t)$ are constant! It suffices to rewrite the constant k as $k = \sqrt{\kappa \Lambda/3}$, then we have $\varrho(t) = \Lambda/c^2$ and $p(t) = -\Lambda$. This corresponds to the original equations (2.6) and (2.7) with $\varrho(t) = p(t) = 0$, $\Lambda \neq 0$, and the solution for $a(t)$ corresponds to that in (2.20).

2.2 Let us assume that we have $7n$ protons and n neutrons at our disposal. (n is an arbitrary integer.) According to the assumption, all neutrons are used up in helium nuclei containing 2 neutrons; correspondingly $n/2$ helium nuclei are produced. However, these contain also 2 protons each, hence n protons disappear in helium nuclei and $6n$ protons (= hydrogen nuclei) are left over. Hence the ratio of hydrogen to helium nuclei is $Z_H : Z_{He} = 6n : (n/2) = 12 : 1$. However, we were asked for the ratio of the densities. A helium nucleus is about 4 times as heavy as a hydrogen nucleus, thus the ratio of densities is $\varrho_H : \varrho_{He} \simeq 3 : 1$ (i.e., 75% H, 25% He).

Chapter 3

3.1 We substitute the expressions from (3.7) for $\Delta t'$ and $\Delta x'$ in (3.8), and obtain

$$(\Delta \tau')^2 = (\Delta t')^2 - \frac{1}{c^2}(\Delta x')^2$$

$$= \gamma^2 \left(\Delta t^2 - 2\frac{v_x}{c^2}\Delta t \Delta x + \frac{v_x^2}{c^2}\Delta x^2 \right)$$

$$- \frac{\gamma^2}{c^2}\left(\Delta x^2 - 2v_x \Delta t \Delta x + v_x^2 \Delta t^2 \right)$$

$$= \Delta t^2 \gamma^2 \left(1 - \frac{v_x^2}{c^2}\right) - \frac{\Delta x^2}{c^2} \gamma^2 \left(1 - \frac{v_x^2}{c^2}\right)$$

$$= \Delta t^2 - \frac{1}{c^2} \Delta x^2 = (\Delta \tau)^2, \tag{A.8}$$

using the definition of γ in (3.7).

3.2 From the formula (3.44) we obtain

$$r_S \simeq 1.5 \times 10^{-27} \, \text{m} \sim 10^{-17} r_{\text{atom}} \sim 10^{-12} r_{\text{nucleus}}. \tag{A.9}$$

Chapter 4

4.1 Instead of (4.4) we now obtain

$$\left(-\omega^2 + c^2 k^2 + \frac{m^2 c^4}{\hbar^2}\right) \Phi(x, t) = 0, \tag{A.10}$$

which is satisfied for all x and t only for $\omega^2 = c^2 k^2 + m^2 c^4 / \hbar^2$.

4.2 First we compute

$$\frac{\partial}{\partial x} \Phi(r) = \frac{\partial r}{\partial x} \frac{\partial}{\partial r} \Phi(r) = \left(-\frac{x}{r^3} - \frac{\lambda x}{r^2}\right) \Phi_0 e^{-\lambda r}, \tag{A.11}$$

then

$$\frac{\partial^2}{\partial x^2} \Phi(r) = \left(-\frac{\lambda}{r^2} + \frac{\lambda^2 x^2 - 1}{r^3} + \frac{3\lambda x^2}{r^4} + \frac{3x^2}{r^5}\right) \Phi_0 e^{-\lambda r}. \tag{A.12}$$

After a similar calculation with $x \to y$ and $x \to z$, it follows that

$$\left(\frac{\partial^2}{\partial x^2} + \frac{\partial^2}{\partial y^2} + \frac{\partial^2}{\partial z^2}\right) \Phi(r) = \frac{\lambda^2}{r} \Phi_0 e^{-\lambda r} = \lambda^2 \Phi(r). \tag{A.13}$$

Hence (4.11) is satisfied for $\lambda = mc/\hbar$.

Chapter 5

5.1 The conservation of total momentum in electron–electron scattering implies

$$\vec{p}_1^{\,a} + \vec{p}_2^{\,a} = \vec{p}_1^{\,b} + \vec{p}_2^{\,b}, \tag{A.14}$$

and the conservation of total energy implies

$$E_1^a + E_2^a = E_1^b + E_2^b. \tag{A.15}$$

With $\vec{p}_2{}^a = -\vec{p}_1{}^a$ it follows from (A.14) that $\vec{p}_2{}^b = -\vec{p}_1{}^b$. Using $E = \sqrt{m_e^2 c^4 + \vec{p}^2 c^2}$ and replacing $\vec{p}_2{}^{a,b}$ by $-\vec{p}_1{}^{a,b}$, it follows from (A.15) that

$$2\sqrt{m_e^2 c^4 + \vec{p}_1{}^{a2} c^2} = 2\sqrt{m_e^2 c^4 + \vec{p}_1{}^{b2} c^2}, \tag{A.16}$$

hence $\left|\vec{p}_1{}^b\right| = \left|\vec{p}_1{}^a\right|$ and also $\left|\vec{p}_2{}^b\right| = \left|\vec{p}_1{}^a\right|$, $E_1^b = E_1^a$, and $E_2^b = E_1^a = E_2^a$.

5.2 From (5.41) we deduce $E_{tot}(n = 2) = -\frac{1}{4}E_R$, $E_{tot}(n = 1) = -E_R$, hence the energy of the emitted photon is

$$E_{phot} = E_{tot}(n = 2) - E_{tot}(n = 1) = \frac{3}{4}E_R. \tag{A.17}$$

In order to calculate the Rydberg energy, we express it first in terms of the fine structure constant α (see 5.31): $E_R = (m_e/2)\alpha^2 c^2$. Thus we have

$$E_{phot} = \frac{3}{8}\alpha^2 m_e c^2 \simeq 10.2\,eV = 16.3 \times 10^{-19}\,J. \tag{A.18}$$

The wavelength of the photon follows from (4.7) and (4.8):

$$\lambda = \frac{c}{\nu} = \frac{hc}{E} \sim 1.22 \times 10^{-7}\,m = 122\,nm, \tag{A.19}$$

corresponding to ultraviolet radiation.

Chapter 6

6.1 We already know the quark content of neutrons and protons, and use the u, d, and s quark masses and charges from Table 6.1, according to which the s quark is about $0.2\,GeV/c^2$ heavier than the u and d quarks. The baryons Λ^0 and Σ^+, Σ^0, Σ^- are 0.18–$0.26\,GeV/c^2$ heavier than a neutron or a proton, and contain, accordingly, one s quark. The nature of the two additional quarks follows from the electric charges, with the result:

Λ^0 and $\Sigma^0 \sim$ (uds), $\Sigma^+ \sim$ (uus), $\Sigma^- \sim$ (dds).

The baryons Ξ^0 and Ξ^- are about $0.38\,GeV/c^2$ heavier than a neutron or a proton, and contain accordingly two s quarks:

$\Xi^0 \sim$ (uss), $\Xi^- \sim$ (dss).

Chapter 7

7.1 An $\bar{s}$ quark with charge $+\frac{1}{3}e$ can turn into a $\bar{u}$ quark by the emission of a virtual W^+ boson. The virtual W^+ boson can decay into the following quarks or leptons (with masses smaller than the $\bar{s}$ mass): $(u\bar{d})$, $(e^+\nu_e)$, $(\mu^+\nu_\mu)$. Thus we obtain the three possibilities $\bar{s} \to \bar{u} + u + \bar{d}$, $\bar{s} \to \bar{u} + e^+ + \nu_e$, $\bar{s} \to \bar{u} + \mu^+ + \nu_\mu$. (Even though the sum of the quark masses $\bar{u}, u$, and $\bar{d}$ is larger than the $\bar{s}$ mass, these quarks can subsequently form relatively light pions.)

7.2 First, the three decay possibilities of the $\bar{s}$ quark lead to the following three decay possibilities of a K^+ meson consisting of a $u\bar{s}$ pair:

$$K^+ \to u + \bar{u} + u + \bar{d} \to \pi^0 + \pi^+, \tag{A.20}$$

$$K^+ \to u + \bar{u} + e^+ + \nu_e \to \pi^0 + e^+ + \nu_e,$$

$$K^+ \to u + \bar{u} + \mu^+ + \nu_\mu \to \pi^0 + \mu^+ + \nu_\mu. \tag{A.21}$$

In addition, a quark can emit a gluon (or even two gluons), which can decay, in turn, into a $u\bar{u}$ or a $d\bar{d}$ pair, leading to additional pions. (According to (6.10), pions are relatively light.) The eight additional decay possibilities are

$$K^+ \to \pi^0 + \pi^0 + \pi^+, \quad \pi^+ + \pi^+ + \pi^-, \tag{A.22}$$

$$\pi^0 + \pi^0 + e^+ + \nu_e, \quad \pi^+ + \pi^- + e^+ + \nu_e,$$

$$\pi^0 + \pi^0 + \mu^+ + \nu_\mu, \quad \pi^+ + \pi^- + \mu^+ + \nu_\mu,$$

$$\pi^0 + \pi^0 + \pi^0 + e^+ + \nu_e,$$

$$\pi^0 + \pi^+ + \pi^- + e^+ + \nu_e. \tag{A.23}$$

(The latter two of these decays, for which we expect a very low probability, have not yet been detected.)

Finally, the $u\bar{s}$ pair can annihilate into a virtual W^+ boson, which can decay either into a $u\bar{d}$ pair (which, after the emission of gluons, can form the final states $\pi^0\pi^+$, $\pi^0\pi^0\pi^+$ and $\pi^+\pi^+\pi^-$ already listed above), or into purely leptonic $e^+\nu_e$ or $\mu^+\nu_\mu$ pairs. The latter lead to additional possible K^+ decays

$$K^+ \to e^+ + \nu_e, \quad K^+ \to \mu^+ + \nu_\mu. \tag{A.24}$$

The decays (A.20) and (A.22) are denoted as *hadronic decays*, those in (A.24) as *leptonic decays*, and those in (A.21) and (A.23) as *semileptonic*.

Chapter 8

8.1

(a) With $R = 27\,\text{km}/2\pi \simeq 4300$ m we obtain for the coefficient $ce^2/(6\pi\varepsilon_0 R^2)$ in (8.6)

$$\frac{ce^2}{6\pi\varepsilon_0 R^2} \simeq 2.5 \times 10^{-27}\,\text{kg m}^2\,\text{s}^{-3} = 2.5 \times 10^{-27}\,\text{W}. \tag{A.25}$$

For an electron with mass $m_e \simeq 5.11 \times 10^{-4}\,\text{GeV}/c^2$ and an energy $E = 104\,\text{GeV}$ we obtain $E/(m_e c^2) \simeq 2.035 \times 10^5$. Then formula (8.6) gives $P \simeq 4.3 \times 10^{-6}\,\text{W} \simeq 2.7 \times 10^{13}\,\text{eV}$ per second. Hence an electron emits an energy of $2.7 \times 10^{13}\,\text{eV} = 27000\,\text{GeV}$ per second, about 260 times its total energy— this energy loss has to be compensated by an energy feed by corresponding electric fields.

(b) For a proton with mass $m_p \simeq 0.938\,\text{GeV}/c^2$ and an energy $E = 7 \times 10^3\,\text{GeV}$ we obtain $E/(m_p c^2) \simeq 7.46 \times 10^3$. The factor $ce^2/(6\pi\varepsilon_0 R^2)$ is the same as in (A.25), hence (8.6) gives $P \simeq 7.7 \times 10^{-12}\,\text{W} \simeq 4.8 \times 10^7\,\text{eV}$ per second. Thus a proton radiates an energy of $4.8 \times 10^7\,\text{eV} = 0.048\,\text{GeV}$ per second, only a small fraction of its total energy.

Chapter 9

9.1

(a) Derivation of the hermiticity of A_{ij} from the unitarity of U_{ij}: we have omitted terms of the order A_{ij}^2 in the series expansion $U_{ij} = \delta_{ij} + iA_{ij} + \ldots$, and likewise we can neglect terms of this order in the condition for unitarity:

$$\delta_{ij} = \sum_{j=1}^{N} U_{ij} U_{kj}^* \simeq \sum_{j=1}^{N} (\delta_{ij} + iA_{ij})(\delta_{jk} - iA_{jk}^*)$$

$$\simeq \delta_{ik} + iA_{ik} - iA_{ki}^* + \mathcal{O}(A^2). \tag{A.26}$$

It follows that $A_{ki}^* = A_{ik}$.

(b) Derivation of the vanishing trace of A_{ij} from $\det(U) = 1$: it is most convenient to use the generally valid formula $\log(\det(U)) = \text{Tr}(\log(U))$ together with the series expansion $\log(1 + \varepsilon) \simeq \varepsilon$:

$$1 = \det(U) = e^{\log(\det(U))} = e^{\text{Tr}(\log(U))}$$

$$\simeq e^{\text{Tr}(\log(\delta_{ij} + iA_{ij}))} \simeq e^{\text{Tr}(iA_{ij})}, \tag{A.27}$$

which is satisfied only for $\text{Tr}(A_{ij}) = \sum_{i=1}^{N} A_{ii} = 0$.

9.2 First, complex $N \times N$ matrices contain $2N^2$ real parameters (two for each complex matrix element). The N^2 conditions $A_{ij} = A_{ji}^*$ reduce the number of free parameters to N^2, and the condition of a vanishing trace by one more. Hence we are left with $N^2 - 1$ free real parameters, corresponding to $N^2 - 1$ linearly independent matrices. For $N = 2$ this number is equal to 3, and for $N = 3$ we find 8 independent matrices (accordingly there exist 8 independent gluons).

Chapter 11

11.1 The relation

$$\alpha_{\text{s,measured}}(E) = \frac{\alpha_\text{s}}{1 + b_\text{s}\alpha_\text{s} \ln(\Lambda^2/E^2)} = \frac{1}{b_\text{s} \ln(\Lambda^2_{\text{QCD}}/E^2)} \tag{A.28}$$

can be written in the form

$$\frac{1}{\alpha_\text{s}} + b_\text{s} \ln(\Lambda^2/E^2) = b_\text{s} \ln\left(\Lambda^2_{\text{QCD}}/E^2\right), \tag{A.29}$$

which gives, solved for Λ^2_{QCD},

$$\Lambda^2_{\text{QCD}} = \Lambda^2 e^{1/(\alpha_\text{s} b_\text{s})}. \tag{A.30}$$

From (A.28) we obtain

$$\Lambda^2_{\text{QCD}} = E^2 e^{1/\left(\alpha_{\text{s,measured}}(E) b_\text{s}\right)}. \tag{A.31}$$

This formula is valid for all E. Using (A.28) for $E = E_1$, and substituting (A.31) for Λ^2_{QCD} with $E = E_2$, we obtain after some calculation

$$\alpha_{\text{s,measured}}(E_1) = \frac{\alpha_{\text{s,measured}}(E_2)}{1 + b_\text{s}\alpha_{\text{s,measured}}(E_2) \ln(E_2^2/E_1^2)}. \tag{A.32}$$

From (A.32) we obtain for $E_1 = 22\,\text{GeV}$, $E_2 = 91\,\text{GeV}$ and $\alpha_{\text{s,measured}}$ $(91\,\text{GeV}) \simeq 0.12$ together with $b_\text{s} = -23/12\pi$,

$$\alpha_{\text{s,measured}}(22\,\text{GeV}) \simeq 0.15, \tag{A.33}$$

which agrees with Fig. 11.5 within the error bars.

Chapter 12

12.1 The sought-after mass M satisfies

$$\frac{e^2}{4\pi\varepsilon_0} = GM^2. \tag{A.34}$$

It follows that

$$M \simeq 2 \times 10^{-9}\,\text{kg} \simeq 1.1 \times 10^{18}\,\text{GeV}/c^2 \simeq \frac{1}{2}M_{\text{Planck}}. \tag{A.35}$$

Comment: For elementary particles of mass about the Planck mass, the magnitude of the gravitational force would be of the same order as the magnitude of the forces originating from the interactions of the Standard Model; in this sense, all four interactions would be "unified".

References

1. A. Einstein, *Relativity: The Special and General Theory* (H. Holt and Company, New York, 1920)
2. A. Friedmann, Gen. Relat. Gravitation **31**, 1991 (1999)
3. N.W. Boggess et al., Astrophys. J. **397**, 420 (1992)
4. D.J. Fixsen et al., Astrophys. J. **420**, 445 (1994)
5. A. Riess et al., Astron. J. **516**, 1009 (1998)
6. A. Riess et al., Astron. J. **560**, 49 (2001)
7. S. Perlmutter et al., Int. J. Mod. Phys. A **15S1**, 715 (2000)
8. P.J.E. Peebles, B. Ratra, Rev. Mod. Phys. **75**, 559 (2003)
9. A.H. Guth, *The Inflationary Universe* (Basic Books, New York, 1998)
10. A.D. Linde, *Particle Physics and Inflationary Cosmology* (CRC Press, Boca Raton, 1990)
11. M. Gell-Mann, Phys. Lett. **8**, 214 (1964)
12. P.W. Higgs, Phys. Rev. Lett. **12**, 132 (1964)
13. P.W. Higgs, Phys. Rev. Lett. **13**, 508 (1964)
14. P.W. Higgs, Phys. Rev. **145**, 1156 (1966)
15. F. Englert, F. Brout, Phys. Rev. Lett. **13**, 321 (1964)
16. G.S. Guralnik, C.R. Hagen, T.W.B. Kibble, Phys. Rev. Lett. **13**, 585 (1964)
17. T.W.B. Kibble, Phys. Rev. **155**, 1554 (1967)
18. G. 't Hooft, Nucl. Phys. B **33**, 173 (1971)
19. G. 't Hooft, Nucl. Phys. B **35**, 167 (1971)
20. G. 't Hooft, M. Veltman, Nucl. Phys. B **50**, 318 (1972)
21. BES Collaboration, Phys. Rev. Lett. **88**, 101802 (2002)
22. Crystal Ball Collaboration, SLAC-PUB-5160 (1989)
23. LENA Collaboration, Z. Phys. C **15**, 299 (1982)
24. MD-1/VEPP-4 Collaboration, Z. Phys. C **70**, 31 (1996)
25. Tasso Collaboration, Z. Phys. C **22**, 307 (1984)
26. Tasso Collaboration, Phys. Lett. B **138**, 441 (1984)
27. Jade Collaboration, Phys. Rep. **148**, 67 (1987)
28. Pluto Collaboration, Phys. Rep. **83**, 151 (1982)
29. Mark J Collaboration, Phys. Rev. D **34**, 681 (1986)
30. Cello Collaboration, Phys. Lett. B **183**, 400 (1987)
31. C.N. Yang, R.L. Mills, Phys. Rev. **96**, 191 (1954)
32. H. Fritsch, M. Gell-Mann, H. Leutwyler, Phys. Lett. B **74**, 365 (1973)
33. S. Weinberg, Phys. Rev. **28**, 4482 (1973)
34. S.L. Glashow, Nucl. Phys. B **22**, 579 (1961)
35. J.C. Ward, A. Salam, Phys. Lett. **19**, 168 (1964)

U. Ellwanger, *From the Universe to the Elementary Particles*,
Undergraduate Lecture Notes in Physics, DOI: 10.1007/978-3-642-24375-2,
© Springer-Verlag Berlin Heidelberg 2012

36. A. Salam, in *Elementary Particle Theory*, ed. by N. Svartholm (Almquist and Wiskell, Stockholm, 1968)
37. S. Weinberg, Phys. Rev. Lett. **19**, 1264 (1967)
38. K. Symanzik, Commun. Math. Phys. **18**, 227 (1970)
39. C.G. Callan, Phys. Rev. D **2**, 1541 (1970)
40. D.J. Gross, F. Wilczek, Phys. Rev. Lett. **30**, 1343 (1973)
41. H.D. Politzer, Phys. Rev. Lett. **30**, 1346 (1973)
42. Jade Collaboration, Eur. Phys. J. C **1**, 461 (1998)
43. Jade Collaboration, Phys. Lett. B **459**, 326 (1999)
44. Topaz Collaboration, Phys. Lett. B **313**, 475 (1993)
45. DELPHI Collaboration, Z. Phys. C **73**, 229 (1997)
46. ALEPH Collaboration, Z. Phys. C **73**, 409 (1997)
47. DELPHI Collaboration, Eur. Phys. J. C **14**, 557 (2000)
48. L3 Collaboration, Phys. Lett. B **489**, 65 (2000)
49. OPAL Collaboration, Phys. Lett. B **371**, 137 (1996)
50. OPAL Collaboration, Z. Phys. C **75**, 193 (1997)
51. LEP QCD Annihilations Working Group, https://lepqcd.web.cern.ch/LEPQCD/annihilations/
52. H. Georgi, S.L. Glashow, Phys. Rev. Lett. **32**, 438 (1974)
53. J. Wess, B. Zumino, Phys. Rev. Lett. B **49**, 52 (1974)
54. H.P. Nilles, Phys. Rep. **110**, 1 (1984)
55. Y. Okada, M. Yamaguchi, T. Yanagida, Prog. Theor. Phys. **85**, 1 (1991)
56. J.R. Ellis, G. Ridolfi, F. Zwirner, Phys. Lett. B **257**, 83 (1991)
57. T. Kaluza, Sitzungsber. Preuss. Akad. Wiss. Berlin (Math. Phys.), **966** (1921)
58. O. Klein, Z. Phys. **37**, 895 (1926)
59. J. Polchinski, Phys. Rev. Lett. **75**, 4724 (1995)
60. B. Carter, in Large Number Coincidences and the Anthropic Principle in Cosmology. IAU Symposium 63: Confrontation of Cosmological Theories with Observational Data (Reidel, Dordrecht, Netherlands, 1974), pp. 291–298
61. S. Weinberg, Phys. Rev. Lett. **59**, 2607 (1987)
62. L. Susskind, in The Anthropic Landscape of String Theory, ed. by B. Carr Universe or Multiverse? (Cambridge University Press, Cambridge, 2007), pp. 247–266: arXiv:hep-th/0302219 (2003)

Index

A

Accelerator, 101, 103–107
Ring-, 101, 103–107, 114
ALICE, 106
α radiation, 7–8
Amplitude, 46, 51
Anderson, 10
Angular momentum, 67–68, 70
Anthropic principle, 171–172
Antimatter, 19, 26
Antiparticle, 10, 12, 66, 69, 94
Astroparticle physics, 119–120
ATLAS, 106, 108
Atom, 3–5, 70–71
diameter, 3
mass, 6
nucleus, 3–10
structure, 3–4
Atomic number, 6

B

B factories, 118
Baryon, 6, 10–11, 76
Becquerel, A.H., 7
β function, 147
β radiation, 8–9
Bethe–Weizsäcker formula, 12
Big Bang, 3, 19, 25–26
Black hole, 41–42
BNL, 113
Bohr atomic model, 70–71, 105
Bohr radius, 70
Bohr, N., 71
Boson, 69

W-, 81–87, 132
X-, 154
Z-, 81–86, 113, 133
Brout, R., 88
Bunch, 105, 108

C

Cabibbo–Kobayashi–Maskawa matrix, 83
Calorimeter, 108
Cartesian coordinate system, 38
Cavity, 105
CERN, 85, 91, 104
Chadwick, J., 6
Charge conjugation, 94
Charpak, 107
CMS, 106
COBE satellite, 21
Color (of the strong interaction), 73–80, 112
Compact space, dimension, 165–168, 170
Complex field, 124–125
Complex number, 123–124
Confinement, 76, 79
Cosmic background radiation, 20–21, 24–25
Cosmic radiation, 96, 119
Cosmological constant, 17, 26, 92, 171
Coulomb solution, 50
Coupling constant, 63, 89, 141, 143, 145–149, 151–155
Covariant derivative, 128
CP transformation, 94
CP violation, 94
Cronin, J.W., 94
Curie, Marie, 7
Curie, Pierre, 7

U. Ellwanger, *From the Universe to the Elementary Particles*,
Undergraduate Lecture Notes in Physics, DOI: 10.1007/978-3-642-24375-2,
© Springer-Verlag Berlin Heidelberg 2012